Fertility of the Sea

Fertility of the Sea

Volume 1

Edited by

JOHN D. COSTLOW, Jr.

Duke University, Marine Laboratory
Beaufort, North Carolina

GORDON AND BREACH SCIENCE PUBLISHERS

New York · London · Paris

Copyright © 1971 by

Gordon and Breach Science Publishers Inc.
440 Park Avenue South
New York, N.Y. 10016

Editorial office for the United Kingdom

Gordon and Breach Science Publishers Ltd.
12 Bloomsbury Way
London W.C.1

Editorial office for France

Gordon & Breach
7-9 rue Emile Dubois
Paris 14e

Library of Congress catalog card number: 74-132383. ISBN: 0677 14960 3. All rights
reserved. No part of this book may be reproduced or utilized in any form or by any
means, electronic or mechanical, including photocopying, recording, or by any information storage and retrieval system, without permission in writing from the publishers.
Printed in east Germany.

Foreword

The Symposium on Fertility of the Sea, the ninth in a continuing series of symposia sponsored by Latin American universities, was held in Sao Paulo, Brazil, December 1–6, 1969, under the Chairmanship of Dra. Marta Vannucci of the Oceanographic Institute of the University of Sao Paulo. This was an important symposium of great scientific and social significance and brought together many scientists from all over the world, some from as far as India. This was made possible through the generous support of international agencies such as UNESCO, FAO, The Royal Society, and foreign governments. Excellent cooperation from the U. S. Atomic Energy Commission and the National Science Foundation helped to ensure the success of the meeting. The generous support of the Ford Foundation through a grant to the National Academy of Sciences is gratefully acknowledged. As in previous symposia, the Oak Ridge National Laboratory was a co-sponsor. The Marine Laboratory of Duke University was the U. S. university co-sponsor.

A symposium on oceanography was originally suggested by Dr. K. N. Rao of the Ford Foundation, and was developed through discussions with Dra. Vannucci and with Drs. Karl Wilbur and John Costlow of Duke University.

The program was arranged to present a picture of the current status of studies on fertility of the sea, a subject of great economic and social importance. The excellent cooperation of Brazilian authorities, from both the National Research Council and the University of Sao Paulo, and the invitation to hold the meeting in the beautiful new Architectural School, left a very good impression on all the visitors and participants.

A sufficient number of important questions were brought up during the course of the meeting to necessitate additional sessions on Saturday afternoon after the official close of the symposium. During these Saturday afternoon sessions a workshop was planned for the following week. Emphasized during this workshop was the need for the development of a number of tropical oceanography stations in the Southern Americas, and the concomitant need of centers for training young people in tropical oceanography to staff these future stations.

The writer, while not in the field of oceanography, was most pleased with the excellent cooperation and the international character of this meeting. The published proceedings should form a milestone in the area of oceanography.

Again, we want to thank out Latin American colleagues for sponsoring the symposium and for making it a most remarkable meeting.

Further symposia in this series are as follows:

1961

International Symposium on Tissue Transplantation, Santiago, Viña del Mar, and Valparaiso, Chile (published in 1962 by the University of Chile Press, Santiago; Edited by A. P. Cristoffanini and Gustavo Hoecker; 269 pp.).

1962

Symposium on Mammalian Cytogenetics and Related Problems in Radiobiology, Sao Paulo and Rio de Janeiro, Brazil (published in 1964 by The MacMillan Company, New York, under arrangement with Pergamon Press, Ltd., Oxford; Edited by C. Pavan, C. Chagas, O. Frota-Pessoa, and L. R. Caldas; 427 pp.).

1963

International Symposium on Control of Cell Division and the Induction of Cancer, Lima, Peru, and Cali, Columbia (published in 1964 by the U. S. Department of Health, Education, and Welfare as National Cancer Institute Monograph 14; Edited by C. C. Congdon and Pablo Mori-Chavez; 403 pp.).

1964

International Symposium on Genes and Chromosomes, Structure and Function, Buenos Aires, Argentina (published in 1965 by the U. S. Department of Health, Education, and Welfare as National Cancer Institute Monograph 18; Edited by J. I. Valencia and Rhoda F. Grell, with the cooperation of Ruby Marie Valencia; 354 pp.).

1965

International Symposium on the Nucleolus—Its Structure and Function, Montevideo, Uruguay (published in 1966 by the U. S. Department of Health, Education, and Welfare as National Cancer Institute Monograph 23; Edited by W. S. Vincent and O. L. Miller, Jr.; 610 pp.).

1966

International Symposium on Enzymatic Aspects of Metabolic Regulation, Mexico City, Mexico (published in 1967 by the U. S. Department of Health, Education, and Welfare as National Cancer Institute Monograph 27; Edited by M. P. Stulberg; 343 pp.).

1967

International Symposium on Basic Mechanisms in Photochemistry and Photobiology, Caracas, Venezuela (published in 1968 by Pergamon Press as Volume 7, Number 6, *Photochemistry and Photobiology;* Edited by James W. Longworth; 326 pp.).

1968

International Symposium on Nuclear Physiology and Differentiation, Belo Horizonte, Minas Gerais, Brazil (published in 1969 by the Genetics Society of America as a supplement to *Genetics,* Volume 61, No. 1, Edited by Robert P. Wagner; 469 pp.).

ALEXANDER HOLLAENDER

Biology Division,
Oak Ridge National Laboratory*
Oak Ridge, Tennessee, U.S.A.

---

* Operated by Union Carbide Corporation for the U. S. Atomic Energy Commission.

Contents

Contents

Synthesis*

J. E. G. RAYMONT

To attempt a synthesis of the papers to which we have listened (well over 30) on various aspects of the fertility of the sea—to do justice to the many excellent contributions—is an impossible task, probably for anyone and certainly for me. May I apologise at the outset that I cannot even mention all the contributions in the brief time available; my apologies also to those authors of papers and equally to those helping in the discussions whom I do not quote by name.

Someone this morning suggested that the methods used by biologists in joint oceanographic investigations compared rather unfavourably with the methods used by physical oceanographers. While, speaking as a biologist, I would agree with this to some extent, I would like to put the matter in possibly a more positive way. I would pay tribute to the physical oceanographers and say how much they have contributed in my opinion to the success of this Symposium.

Although this is a platitude, we must stress at the outset that productivity turns ultimately on primary production. The whole fertility of the sea must depend upon primary production. Everyone knows this, and however much you may accept my devious discussions of this morning dealing with alternative pathways in the marine ecosystem, we are forced back ultimately in space and time to primary production. Now the difficulty which was emphasized quite early in Dr. Lafond's paper was that primary production is

---

* The synthesis was originally to have been recorded. Owing to a power breakdown about half the summary was given without notes and has been prepared largely from memory. This, however, has been added to the original recording and only minor editing and corrections have been made in order to maintain a general discursive style throughout.

usually held back by nutrient supply. Of course production is limited by light ultimately; except in very very shallow inshore waters this must be the ultimate parameter which limits photosynthesis. But whatever the depth of the productive photosynthetic zone, which of course depends essentially on light penetration, we are soon faced almost everywhere with the problem of how to get the mass of nutrients which lies below the thermocline into the euphotic productive zone. At this point, therefore, the problem of the thermocline and how far this is a barrier to upward passage of solutes becomes paramount.

I would like to begin, therefore, with a brief outline of some of the investigations discussed in this Symposium dealing with physical parameters.

Early on, Dr. Lafond showed us the way in which the thermocline acted as a barrier. The problem of upward transport of solutes seemed at first insurmountable in areas of strong permanent thermoclines, but Dr. Lafond illustrated eddy formation in the oceans. This formation of eddies, except on a very large scale, could only be demonstrated by intensive and repeated surveys of particular areas. There was also an excellent discussion of the thermocline; some of us had perhaps naively felt this was a more or less continuous constant and homogeneous sort of barrier. But it could have a finer structure and could be divisible into micro layers. To some extent, therefore, it could permit the upward passage of solutes. We next proceeded to a review of the major ocean currents and water masses, and our attention was focussed on the great western boundary currents with the very large eddies which they produce. Thus we were introduced to advective and turbulent mixing. Dr. Capurro then drew our attention to the wide spectrum of eddy size. This, I think, is of the utmost importance; not only are there major eddies, but where the physical oceanographers have been able to examine certain areas in detail, much smaller fine eddies also appear at least in particular places. Internal waves, which in all conscience are difficult to analyse and quantify, also contribute to mixing. Mixing is the one factor which we need as it were for promoting fertility. But there is also thermodynamic mixing; in what appear to be stable layers, because of the differential rates of temperature and salinity flux, even in the thermocline there can be instability and this, of course, assists in mixing.

So we came to upwelling which seems to have been one of the major topics of discussion in the Symposium. But before we discuss particular regions of upwelling, I would like to draw attention to someone's quotation of a calculation from Munk. This showed that over the Pacific Ocean as a whole there was vertical upward passage of something approaching 1.2 cm water per day. It has been suggested during the discussion, with all due respect to Munk, that this figure may have to be modified in the future, but in any case the very fact that there exists this general upward passage of water seems to me overwhelmingly important; nutrients may be carried up to some extent over the whole of an ocean into the productive euphotic zone.

Then we came to upwelling in the stricter sense. As I asked in discussion (and I am still not sure that I got an answer) "What is upwelling?" We all seem to have our own varieties; there is major upwelling and minor up- welling. But one thing is absolutely clear; the importance of upwelling in the fertility of the sea is outstanding.

We have in a paper from Dr. da Silva the suggestion that in southern Brazil off Cabo Frio there was a fairly regular swing round, as I would term it, of wind direction. This could be over relatively short periods and lead to relatively short term upwellings of the order of a week. However, part of the cycle of changing wind direction may prevail for longer periods, and this might lead to seasonal upwelling. Some discussion followed on the time factor in upwelling and in particular the age of upwelled water, and there was argument as to whether upwelling could persist over long periods. There was then an analysis of what appeared to be perhaps more major upward movement of water in the same area off Cabo Frio. This, I think, was Dr. Rock's contribution. He showed that during the course of the Brazil Current, off the shelf area, a great eddy occurred but that further complications were produced in the flow of water around that area due to bottom topography. Rock demonstrated the existence of a submarine peak and the effect that this had on the vertical stratification; there was a very marked surface temperature-salinity anomaly for the area. The great eddy and submarine peak together then helped in upwelling and general mixing. Rock went on to show, as far as I remember, that there were three major hydrographic layers in the area. The warm saline Brazil Current water was sandwiched in between a surface coastal layer and a deeper layer which was moving towards the coast along the bottom of the continental shelf leading to upwelling.

Further, on the subject of upwelling, we had the contribution of the Oregon group (Dr. Smith) with the very beautiful analysis of upwelling off the Oregon coast. What is so pleasing from the physical oceanographers in this connection is the marked attempt to quantify the upwelling process.

1*

Dr. Smith describes the deeper water welling up but being kept close to the coast and then returning after a certain amount of warming up, presumably to go down beneath the more or less permanent pycnocline in this area. The group had been able to obtain some appreciation of the velocity (and thus the quantities) of the horizontal current flowing approximately south, and they had also pointed out the presence of a deeper counter current flowing northwards. It appeared at first that we were going to obtain an almost complete quantified picture here. May I say I was almost glad when Dr. Smith pointed out that we were still lacking any real knowledge of the rates of vertical transport! He emphasised (and I think all of us would echo his remarks) that since upwelling is of such significance in fertility, a real attack on the rate at which vertical transport of water occurs is a tremendously important item which must be tackled.

This type of work on upwelling has been carried over to the coast of Peru, also by the Oregon group, working with researchers from Peru. Once again there is an excellent analysis of upwelling; there is a beautiful correlation between low temperature and high nutrient water. Again, there was a demonstration of horizontal water movement, though this was to the south, that is against the prevailing wind; presumably the current moved north first and then turned round in a southerly direction. Again it was hoped that in the future some knowledge of the vertical velocity of upwelling would be obtained. An important item in the upwelling work off the Peruvian coast was the patchiness. This phenomenon is of course well known and is reported in all the classical text books on oceanography, but it was interesting to see that with modern methods upwelling was again demonstrated not as a continuous constant process over a whole region but as a series of discrete patches which varied in intensity.

In order to study this type of phenomenon we obviously need synoptic data. Dr. Smith used a monitoring device on a plane to record water temperatures and thus try to obtain a synoptic picture off the Oregon coast. But I would like to refer here particularly to the paper by Drs. Mascarenhas and Almeida who described in some detail the actual remote sensing device which can be used on a aircraft. Sensors covering the visible region of the spectrum were most useful in noting such features as currents, river outflows and current boundaries. In connection with our present study, however, the most beautiful results came from sensing infrared rays, so that one could map the boundary of warm and cold currents, but above all note areas of upwelling, even if these were not of great intensity. This need for a synoptic picture of a large area is surely essential, especially if we are trying to study upwelling where it does not occur over a vast area with great intensity, but occurs only in patches. I would like to mention also in this context the device described by Dr. Ballester which as far as I could follow (I apologise for my inadequate knowledge of foreign languages!) plots temperature and salinity data rapidly, as well as monitoring other parameters, so that once again this is a technique for obtaining relatively rapid pictures of what is going on over wide areas of water.

So much for my very inadequate coverage of the contributions from the physical oceanographers, and I apologise for having left out so much! However a question from the physical oceanographers lead on to my second major item for discussion: "How far should air/sea interaction studies (meteorological work) have preceded our investigations of water currents, water masses and so forth?" One of the things which seems to me to have emerged during the course of this Symposium (though perhaps we were not able to devote a great deal of discussion to it) was the need for studies of air/sea interaction. I think we all appreciate that a very full study of meteorological data is absolutely essential. Only then may we attempt to have a better understanding of the reasons for variations in current systems, and even to try to predict them in the long term. Variations in the current systems certainly occur as we are all fully aware.

A particularly important contribution here came of course from Dr. Taft's discussion of monsoon effects. Without attempting to go into this paper in detail, I recall some of the major effects of the changing monsoon pattern, especially in the Indian Ocean area, and the great upwelling which occurred off the Somali coast. The effect was so strong that the thermocline literally breaks the surface. Although, as Dr. Oren described this morning, this may lead to mass mortality of fishes, I think that we cannot count this as too strong an argument against the generally tremendous positive effect which upwelling has in promoting fertility. Many other points came from Dr. Taft's paper; for example, with the changing of the monsoon pattern the matching of wind and current may be very much poorer in some areas than others, for instance, off India. One thing, however, which appeared of particular interest was the demonstration of what appeared to be major vortices off the Somali coast of very considerable extent; these gradually faded out towards the east. Dr. Taft pointed out that these ideas arose partly from theoretical reasoning rather than from observed fact. Nevertheless, if there are these great discrete "pockets" with their own particular current patterns, the mixing processes in this area of the Indian Ocean must be very different from that in similar latitudes in the Pacific or Atlantic Oceans, and such mixings must also have tremendous repercussions on phenomena including fertility.

Having said something then about physical parameters and a little about meteorology, we come particularly to the effects of upwelling and also to the somewhat less dramatic effects of minor mixing. In order to get some plan into the many effects of water movements, we might first concentrate on major nutrients, that is to say on nitrates and phosphates.

One sees particularly in the kind of waters we have here in parts of South America the enormous differences due to upwelling on any scale. Several contributions from workers off the Peruvian coast showed, for example, a very good correlation between strong upwelling regions, as determined by physical analyses, and areas of rich nutrients. These patches of rich nutrients were interspersed between waters of considerably lower nutrient concentration. Generally in the tropics, as is well known, surface or near surface waters are relatively poor in nutrients. In general Dr. Guillen and his associates showed high levels of nutrient concentration in certain areas all fairly close along the coast and mostly associated with the Coastal Peruvian Current. But the areas shifted from season to season with variations in the circulation but above all with variations in upwelling intensity. Dr. Smith also pointed out that off Peru analysis showed that regions with the highest nutrients occurred close to shore in areas marked by the coldest waters where upwelling was at a maximum. Dr. Small, dealing with upwelling off the Oregon coast, also demonstrated an excellent correlation between high nitrate values and intense upwelling areas near the coast. He points out that this rich nitrogen water usually does not spread out to sea, owing to its being confined by relatively warm nutrient-poor water extending southward. Occasionally, however, water masses, rich in nitrate, may be isolated nearer the surface if the pycnocline breaks down temporarily. Similarly in the Caribbean, an analysis showed that areas occurred off the Venezuelan coast of upwelling near the coast followed by down welling of water at a greater distance from shore. In all these varied areas we see a great increase in primary nutrients coming near the surface when upwelling is intense.

This discussion of the primary nutrients in upwelling regions may lead us to a discussion of Dr. Postma's paper which dealt more particularly with the major exchange of nutrients on the grand scale throughout an ocean. The changes in phosphate concentration both in depth and laterally in an ocean were discussed, and though it appears that we have considerable knowledge of upwelling and other processes, the rates of supply of a primary nutrient such as phosphate are largely still unknown. Among other factors, we want to know the rates of transport of nutrients through organisms from the euphotic zone to the deeper layers and also the rates of liberation of phosphate from detritus and other sources. Calculations given included the contributions of phosphate from Antarctic deep bottom water and also the calculated movement upwards of nutrient into the upper layers. During discussion, one of the problems which emerged is the turnover time of a nutrient such as phosphate, and also how much of the material settles in such deep water that at least temporarily it is "lost" for a period.

During discussion the problem also emerged of contributions from drainage from the land. Perhaps we have not considered yet the full significance and the tremendous effect, particularly in tropical environments, of coastal drainage on the fertility of the immediate shallow waters in the area. I am thinking at the moment only of primary nutrients but there is little doubt (as I think Dr. Goldberg pointed out, although he suggested that he had not found good quantitative data to support the argument) that the amount of nitrogen and phosphorus generally in river water is relatively high. This, of course, may lead to increased production in the shallow waters near the river drainage area. Here, however, we came up against a contrast between the Amazon and the Columbia River. The River Amazon waters are relatively low in phosphate and not even very abundant in nitrate, though relatively high in silicon. One cannot assume, therefore, that coastal drainage is going to lead inevitably to enrichment. In spite of these exceptions however, it seems a fairly general rule that river outflow will usually lead to higher nutrient levels. This phenomenon clearly needs careful study as Dr. Goldberg emphasized. Coastal areas on the whole tend to be more productive areas, and after all are much more accessible and usually are much more fishable areas. In brackish water areas and estuaries particularly we cannot therefore just accept the conclusions which have been drawn from studies in the open seas. Hydrography and nutrient cycles demand study and we cannot even use the same methods necessarily as in the open sea. This discussion relates to another argument which emerged this morning; lagoons have their own particular problems and these must be investigated.

I would now like to leave our examination of upwelling in bringing major nutrients (i.e. nitrates and phosphates) to the surface and, although I have dealt with the many problems only hastily, to turn our attention to nutrients other than nitrates and phosphates.

Perhaps we might begin with trace metals. We have already spoken about the generally high productivity of coastal waters. Coastal waters often tend to show many times the productivity of oceanic waters at the same latitude; Mr. Savage mentioned in discussion a factor of × 15 in one temperate area. But there are exceptions, and these as Dr. Goldberg mentioned are largely due to the fact that river water is so variable in its composition. Despite this variability, Dr. Goldberg showed that several trace metals are in much higher concentrations in river waters, and several papers have referred to the question as to how far these greater quantities of trace metals brought down by the rivers might be responsible for increased inshore production. Not all the trace materials reach the sea; there may be problems of adsorption on to other minerals and on organic particles in the estuaries. Nevertheless, the analysis of trace metals and their relative amounts in river waters, coastal waters and the open sea is an area of study which seems to us to warrant intense investigation.

There is no doubt about the complexity of the problem, especially as to how far trace metals, chelators, and similar substances affect primary production. In this connection I would refer particularly to Dr. Smayda's paper. He questioned what were the substances essential for production; he suggested that we had tended to move in our thinking with our techniques, starting with the major nutrients, proceeding to such trace elements as iron, and then progressing to chelators, vitamins, and so forth. In his experiments he had selected four species of phytoplankton for bioassay. He tested water collected at different depths, particularly from the Sargasso, which had been filtered but was then added to the bioassay cultures and also to this were added different mixes of various nutrients. Some of his results were unexpected. For instance, water collected at 100 metres from the Sargasso was "bad" and might actually cause mortality of the algae, though this was eliminated by addition of nitrate, phosphate, silicon and iron. Vitamins were also very effective. On the other hand, surface waters did not appear to require the "normal" nutrients; these could even depress growth, but surface water required above all trace metals and vitamins. In more or less enclosed areas off Puerto Rico, Smayda obtained again very different results on enrichment dependent on species, and this applied to some extent to Caribbean waters also. Smayda suggested that it was impossible to predict for any particular phytoplankton species, or for any particular area or water mass, *exactly* what nutrient mixture was necessary. Although there was much discussion on how far one could use laboratory cultures for bioassay work of this type, the paper was invaluable in demonstrating that different types of water and different species have different nutrient requirements.

This paper brought into sharp focus the whole problem of total enrichment effects of land drainage on phytoplankton production. I cannot help being reminded (indeed some contributors have already referred to the fact) that many years ago the possible positive effects of land drainage were spoken of, certainly by Gran and by Bigelow, but it seems only now are we beginning to attack and quantify some of these factors.

Again somewhat in this context we had the contribution from Dr. Provasoli referring to his demonstration many years ago of the importance of vitamins, chelators and organic substances as a whole on primary production. Vitamins, especially B 1, B 12, and biotin, are now generally accepted as of great importance. But a whole range of substances not yet fully identified and produced by a variety of organisms (seaweeds, bacteria, etc.) all may have positive or negative growth effects. Some may affect the type of growth; moreover, certain species of phytoplankton, while not having an absolute requirement for these vitamins, may vary their composition to some extent when vitamins are abundant. The long term effect on fertility of the mass of ectocrines produced in sea water with their multiple effects including the "quality" of food organisms is outstanding.

Related to this organic enrichment, we have papers such as those of Dr. Prakash, dealing with the effects of humic substances derived from the land. Although humic substances are extremely complicated chemically, he showed that these substances could not only act directly as chelators, but could extend the active growth period of certain phytoplankton species. There were also certain fractions of humic substances, in particular those of low molecular weight, which were especially potent. So the whole problem of organic enrichment must demand more of our attention in the coming years. Despite the criticisms which have been levelled from time to time against using laboratory cultures, surely, if we are going to tackle problems of this sort, the first stage must be an analytical approach, attempting to work out the effects of chelators, trace metals and other substances on cultures. Even though we accept that such laboratory cultures may have suffered adaptation and change and may be very different from natural phytoplankton populations, we must start in the laboratory. There are, of course, other difficulties in laboratory culture techniques. An example is that demonstrated by Dr. Davies in his work on the growth of flagellates in the presence of iron at limiting concentrations. Under normal laboratory conditions much of the ferric hydroxide appears to stick to the wall of the culture vessel. Rate of detachment thus becomes an important factor in growth. There was also evidence that the minimum iron requirement changed as the culture became adapted.

I would like to raise a point here which several contributors made, namely, that the sites of intense upwelling and of high concentrations of primary nutrients are not necessarily the sites of high phytoplankton production. Indeed very often the reverse is the case. Dr. Barber suggested that deeper upwelled water often did not show an immediate growth of algae if the inoculation of cells was small, or without the addition of a mix of trace metals. There could be a long lag phase in growth, but here one could differentiate between what appeared to be "good" water, production being reduced by charcoal treatment, whereas the same treatment had no effect on "bad" water. To either, however, the addition of trace metals and/or chelator such as EDTA produced rich "normal" water, well conditioned, and giving good growth. Barber believes that upwelled water which is reasonably productive has already been conditioned by the growth of phytoplankton itself which supplies the necessary chelators—substances analogous to EDTA. Growth will occur rapidly, and the quantity of trace metals, though always low, is sufficient to allow full growth.

Regarding a possible lag phase in phytoplankton growth in recently upwelled water, we might question to what extent "seeding" as I might term it of an area with phytoplankton cells is required. Or, as Dr. Barber has suggested, are more subtle influences involved and does at least some upwelled water need to be conditioned before it can raise a phytoplankton crop?

I come next to the phytoplankton itself though of course it is artificial to separate the problem of nutrients of all kinds from the phytoplankton crop; indeed, we have just been discussing both to some extent. If however we may now concentrate on the phytoplankton, we have seen how primary production, measured chiefly by the C^{14} method, and estimates of phytoplankton crop coincide fairly well. There is the exception to which I have just alluded—often recently upwelled water may not show a rich crop, but production may lag also. Nevertheless, over broad areas, zones of general upwelling do correspond to zones of high production. This appears to apply particularly in latitudes such as those with good light conditions; any increase in nutrients tends to be followed both by high production and high crop. This was well shown by the work of Dr. Guillen and his associates in the very rich coastal waters off Peru, where production and crop corresponded excellently both in time and with the areas of nutrient rich water. Intrusions of poor tropical waters were marked by low production. However, the dominant species of phytoplankton showed some variation with season; it was especially good that included in the work of the Peruvian group was an analysis of phytoplankton species typical of the different water masses. In this connection (though I do not speak as a systematist) I would echo the plea made by someone else this morning for an urgent attack on systematics in plankton studies. We appear always to have used measurements of chlorophyll *a* as an estimate of crop though the Oregon workers have used C/N ratios also. It is worth noting incidentally that this same group also found an alternation of actively phytosynthesing sites and high chlorophyll with areas of intense upwelling and high nutrients. Nevertheless, in this investigation also we have the same general correspondence that the major regions of upwelling as a whole coincide with major areas of production.

I would like to return briefly, however, to the topic of systematics in phytoplankton studies. The whole problem of dominant species, species diversity and delimitation of communities is one needing urgent attention. Dr. Ferguson Wood discussed this problem and pointed out the need of modern computerised methods in this sort of work. Related to this problem also is the fractionation of phytoplankton and the relative contribution made by various size groups to crop and production (cf. the paper by Mr. Savage). Dr. Tundisi's studies emphasised the importance of nanoplankton in tropical waters, but in inshore areas somewhat larger phytoplankton may flourish at least at times when rainfall and land drainage increased eutrophication.

A problem which emerged again was the proper assessment of living actively photosynthesing cells. One of the most difficult problems is with deep living phytoplankton, well below the euphotic zone. Dr. Holm-Hansen's description of his very beautiful ATP method points the way to the development of modern techniques for the analysis of living material, both deep living phytoplankton and bacteria. But again the problem emerged of the *quality* of the material. The determination of crop as biomass or even as organic content (in the few cases where this has been accomplished) may not be sufficient. The precise quality of the food material as Dr. Provasoli emphasised may be of profound significance in the growth especially of the secondary producers.

Although we have referred several times in passing to light, this parameter has not been dealt with in any detail. Generally the light conditions in tropical and sub-tropical areas are of course ideal for production, but the turbidity of inshore areas, especially mangrove swamp, may seriously reduce light intensity and thus production. Dr. Goldberg also drew our attention to the effect in estuaries where the high silt load may greatly reduce light penetration, a point emphasized by Mr. Savage in discussion on the reduced thickness of the euphotic zone. In the Antarctic it was interesting to learn that in areas of intense phytoplankton blooms, especially for instance in coastal waters, the shading effects of algal crop could reduce light intensity greatly. Dr. Panikkar referred also to the high turbidity on the west coast of India. One important paper on light was that of Dr. Jitts, who pointed out the serious errors which can creep into our measurements of light penetration when the usual type of photocells are used to measure light intensity. The problems of providing suitable illumination in simulated C^{14} methods of measuring production were also analysed. Some of the difficulties of using C^{14} especially in inshore highly eutrophic waters came under discussion from Dr. Teixeira. It was interesting to see in Dr. Tundisi's studies in the same environment the very strong seasonal cycle in primary production which seems to be largely associated with rainfall and land drainage.

One of the very interesting papers in this Symposium was that of Dr. Roels and his associates involving plans for nutrient fertilization experiments in the marine environment. There were other papers, for instance, that of Dr. Schmitt, involving energy budget considerations on nutrient enrichment and animal protein production. Dr. Roels' proposal was to pump seawater from off St. Croix Island where deep oceanic water occurs only a mile or so offshore. The nutrient content of waters below a few hundred metres depth is, of course, relatively very high. Small scale laboratory experiments on phytoplankton growth enriched with water from different depths had shown the expected positive effect on growth. In the experiment now planned, water is to be pumped from some 800 metres and would be transferred to a tank ashore. There its lower temperature would cause condensation of water at the relatively high humidity of the island. This condensed water would then be used for irrigation purposes, and it had been calculated that a substantial economic contribution would come from refrigeration uses and from the supply of water alone. However, after the seawater in the land tank had reached near ambient temperature, it would be pumped to enclosed lagoon areas where experiments were planned on the growing of phytoplankton.

The whole feasibility of such a development provoked considerable debate since the results may be of particular significance to countries which may need water on the one hand and are poor in protein resources on the other. The question as to why in laboratory experiment water from a particular depth gave somewhat better phytoplankton growth than that from other depths, with apparently similar major nutrient concentration, led again to consideration of the particular quality of the water. Minor constituents which may be necessary for healthy growth were again reviewed. The question of which particular species of algae would be harvested in such a development also came under discussion; the problems of blooms of noxious algae was a possible difficulty. The debate again brought to light the necessity of identifying phytoplankton, and of knowing something of the quality of the product in relation to the growth of secondary producers.

As regards secondary production we had an interesting paper by Dr. Bé on biomass and volume estimates of zooplankton crop in the North and Central Atlantic. A good correlation was demonstrated between areas of high zooplankton crop and high primary production. A contribution by Dr. Tomo also dealt with variations in phytoplankton and in the zooplankton seasonally and also the problem of vertical migration in zooplankton. He dealt with the food chain particularly in the Antarctic including the feeding of whales, birds and leopard seals on zooplankton. On the whole, however, there were rather few papers on secondary production, and further links in the food chain were not as thoroughly covered as earlier aspects of the fertility of the sea. In a sense it is worth emphasising that one can speak of fertility almost at any level in the food chain, though of course fertility must always depend ultimately on the level of primary production. But the need for more intensive surveys of zooplankton and studies of the efficiency of conversion and problems of food chains all must become part of our examination of the fertility in the oceans. Several contributors in discussion pleaded for improved techniques in secondary production.

The benthos must also depend ultimately on primary production. Dr. Rowe emphasised this point in reaffirming the principle that the richness of the bottom fauna declined very markedly with increase in depth. But he also made interesting comparisons between one area and another at constant depth. For example, he demonstrated the comparative poverty of the benthos in the Gulf of Mexico in relation to the North Atlantic and to an area off Peru, and he claimed that this poorness of the bottom fauna was related to differences in primary production. Other papers on benthos included one from Dr. Korringa, who pointed out the relatively direct and efficient conversion of primary production to secondary production by bivalves. He gave datails of some practical applications of shellfish farming and also of the dangers of pollution. Several contributors drew our attention to areas where apparently production is relatively high and which have great concentrations of benthos. For example, Dr. Tommasi spoke of the rich Bay of Ilha Grande which is marked by great banks of ophiuroids. Dr. Jakobi referred to areas of high concentration of microfauna which appeared to be equally rich in shrimp production. Rich shrimp productive areas were also referred to by Dr. Kurian speaking of the mudbanks of S. W. India. He discussed the variations in benthos particularly in relation to the type of substratum.

One is faced in connection with the benthos not only with overall primary production but with food supply other than phytoplankton–detritus and so forth. This problem I have dealt with earlier this morning to some extent in discussing the role of detritus, organic aggregates and bacteria and of possible pathways from dissolved organic matter through bacteria and protozoans to benthos and deep living zooplankton. Though I spoke of these alternative pathways with particular reference to the deep sea, they are equally if not more important in shallow waters and contribute there to the great richness and fertility of inshore areas. A particular example of the great richness and general fertility of inshore areas applies to estuaries where the term "trophic trap" was referred to by some of our contributors. Estuaries often maintain a high level of productivity, partly due to the very rapid recycling through the various trophic levels.

A further link in the food chain in the marine environment leads to fish. Dr. da Silva correlated the changing abundance of sardine catches off the south Brazilian coast with periodic upwelling. Our attention was also drawn to the rich anchovy fisheries off the Chilean and Peruvian coasts and to the richness of phytoplankton and zooplankton crops in the same areas. A particularly interesting discovery here was a variation in the feeding of different races of anchovy, on phytoplankton and zooplankton respectively. Dr. Oren dealt with another type of fish production, namely, the production of mullet which can of course feed directly on phytoplankton and filamentous algae, as well as to a considerable extent on detritus. He emphasised that the high production of mullet in this very direct food chain in the marine environment may be of considerable practical significance.

Again, it is possible to speak of fertility but now in terms of fish production. The discussions at this Symposium have perhaps emphasised the need for applying the term "fertility" almost to any level in the extremely complicated ecosystem of the marine environment. Another general point which has emerged is that fertility is rarely, if ever, attributable to one factor. In the past we have been perhaps too much pre-occupied with a single factor such as phosphate or nitrate, or iron, or perhaps chelators or vitamins. And we cannot study only primary production, or secondary production; we need to study the food chain.

I think we have agreed that some of our methodology in biological oceanography does not have a sufficient degree of sophistication and our data are often not sufficiently quantified. But not only do we need sophisticated methods in any study of the fertility of the oceans; the multi-disciplinary attack is absolutely essential, with contributory and correlated studies from the physical oceanographers, the marine chemists, the biochemists, and the marine biologists—indeed all oceanographers of varied interests. This can probably be achieved only by good collaborative studies between groups of workers in various institutions. Finally, since the fertility of the sea is so complex, extensive studies over a period of years is absolutely essential if our results are to have real meaning.

Plankton abundance
in the North Atlantic Ocean*

ALLAN W. H. BÉ, JOSEPH M. FORNS, and OSWALD A. ROELS

Lamont-Doherty Geological Observatory of
Columbia University
Palisades, New York

Abstract

Geographic variations in zooplankton biomass, expressed as displacement volume, wet weight, dry weight and ash-free dry weight, were determined from 342 plankton samples collected during 1958–1968 over wide regions of the North Atlantic. Our study corroborates previous investigations that zooplankton abundances are high in subarctic and cold-temperate waters, along oceanic margins, upwelling regions and active current systems— where nutrient-rich mixed layers exist for varying durations of the year. Low plankton abundances occur in the central water-mass or Sargasso Sea, with minimum values in the southwestern sector.

The regional variations in zooplankton abundance are of sufficiently large magnitude that they mask variations due to diurnal vertical migration, local patchiness and non-synoptic sampling. However, seasonal variations in high latitudes are very large and they are likely to alter significantly our summer and fall estimates of biomass.

Resumo

Variações geográficas na biomassa de zooplâncton, expressa como volume deslocado, pêso úmido, pêso sêco, e pêso sêco sem cinzas, foram determinados com 342 amestras de plâncton coletadas durante 1958–1968 em grande

* Lamont-Doherty Geological Observatory of Columbia University Contribution No. 0000.

Fertility of the Sea

regiões do Atlântico Norte. Êsse estudo confirma investigações anteriores, de que abundâncy do zooplâncton é alta em águas sub-articas e temperadas-frias ao longo de margens Oceânicas, regiões de ressurgência e sistema de corrente ativos-onde águas ricas em nutri entes ocorrem durante várias épocas do ano.

Baixa abundância de plâncton ocorre nas massas de água centrais do Mar de Sargasso, com valôres mínimos no setor sudoeste.

As variações regionais na abundância do zooplâncton são suficientemente grandes para mascarar variações devidas á migração vertical diurna, "patchness" e amostragem não sinótica. Entretanto, variações sazonais em altas latitudes são muito grandes e podem alterar significantemente a estimativa da biomassa no verão e no outono.

INTRODUCTION

This study is concerned with the distribution of plankton standing stock over wide regions of the North Atlantic. Standing stock of plankton, as measured by its displacement volume, wet weight, dry weight or ash-free dry weight, is the total amount of living biomass contained in a parcel of water at a specific time, locale and depth range. The simplicity and wide-spread custom of determining one or more of these parameters is counteracted by the limited significance of such biomass data.

Standing stock values are only valid as static measurements of instan-taneous conditions. More meaningful are estimates of production rates, but they are difficult to obtain, because they require continuous monitoring of mobile, oceanic populations over a seasonal or annual basis.

A second limitation is the usual, inevitable lumping of phytoplankton and zooplankton in biomass determinations. Generally, however, the bulk of the biomass collected by fine- and medium-meshed nets (i.e. 76 to 500 μ mesh-aperture) in openoceans areas distant from land consists predominantly of microzooplankton. Phytoplankton contributes significantly to the plankton biomass of net hauls in near-shore regions and in higher latitudes during the spring and summer.

Another serious problem is the inclusion of various plankton groups belonging to different trophic levels as a single entity of biomass. Since trophic levels are known to cut across taxonomic lines, we cannot simply rely on enumerating plankton groups to unravel their trophodynamic relationships. In productive, near-shore and upwelling regions, the high density of primary producers and herbivores are largely responsible for a short food chain of $1^1/_2$ to 3 steps (Ryther, 1969). Cushing (1959) has considered the high-latitude region as "unbalanced" because plankton production is largely confined to the summer months. In the central water-masses of the oceans, such as the Sargasso Sea, the relatively low phytoplankton productivity supports smaller concentrations of herbivores and gives rise to a higher proportion of secondary and tertiary consumers. According to Ryther (1969) a food chain consisting of five trophic levels can be expected for the oceanic province, where the carnivores play an increasingly important role.

Our sampling depth range of 0–300 m probably serves to integrate any variations due to vertical migration of microplankters that are several millimeters or less in length. Wiborg (1955, p. 62) noted in his detailed seasonal plankton studies at Station M in the Norwegian Sea that "in hauls taken down to 200 m, diurnal vertical migration does not influence the quantity of plankton to any extent, except for euphausiids which usually avoid the net during the day". Since there is little evidence in the literature that microplankters migrate diurnally *en masse* above or below our maximum sampling depth level of 300 m, we believe that day or night-time sampling does not significantly affect the biomass quantities for our depth range.

In the following sections we shall discuss the geographic variations of plankton biomass in the North Atlantic. The term "biomass" has been restricted to "wet weight" by some investigators (e.g. Kanaeva, 1965; Tranter, 1962) and it connotes "dry weight" to others (e.g. Lovegrove, 1962; Hopkins, 1969a, b). We shall use it in a broad context to include all four common methods of measurements: displacement volume, wet weight, dry weight, and ash-free dry weight.

METHODS AND MATERIAL

The majority of the plankton in this study were collected on research vessels *Vema* and *Conrad* of the Lamont–Doherty Geological Observatory. Additional samples were obtained on U.S. Coast Guard, U.S. Navy and Canadian ships. All plankton samples were collected in a uniform manner with standardized equipment that is essentially identical to the one recommended by the Scientific Committee on Oceanic Research's Working Party 2 (Fraser, 1968). The plankton nets had a 50 cm × 50 cm mouth opening and 202 μ mesh-aperture (Nitex 202). All tows were taken in a vertical or oblique manner in the upper 300 m and were raised at a speed of 30 m/min. A Tsurumi–Seiki Kosakusho flowmeter was fitted into the mouth opening of each net to determine the volume of water filtered. More recently, a second flow-meter was placed outside the net to record unobstructed water flow. The difference between inside and outside flowmeter readings indicated the net's filtration'efficiency and the degree of any clogging.

The tows were taken over many seasons and years and at different times of the day. Hence, diurnal, seasonal and long-term variations need to be considered in interpreting the results. Figure 1 shows the locations of the samples and Table 1 indicates that the majority were collected during the warmer months from May through December.

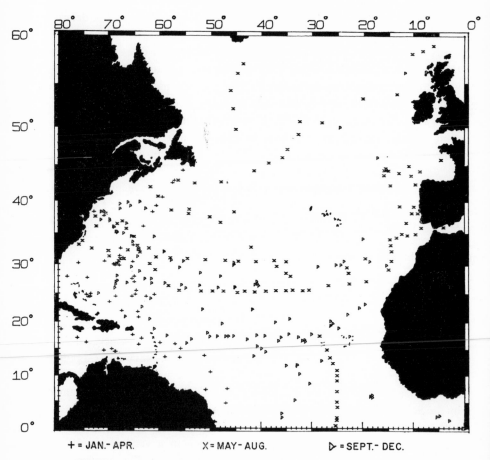

+ = JAN.- APR.   X = MAY - AUG.   ▷ = SEPT.- DEC.

Figure 1 Location of plankton stations and time of collection grouped in four-month intervals

Table 1   Monthly breakdown of 342 plankton samples used in
present study

January	16 samples	July	34 samples
February	17 samples	August	75 samples
March	11 samples	September	19 samples
April	19 samples	October	38 samples
May	20 samples	November	57 samples
June	15 samples	December	21 samples

LABORATORY PROCEDURE

The plankton biomass for 342 samples was determined and expressed in
terms of displacement volume, wet weight, dry weight and ash-free dry
weight. Our laboratory procedure involves a sequence steps that allows the
four measurements to be made with a minimum amount of time and
apparatus. Disproportionately large and gelatinous organisms (> 3 cm),
such as salps, siphonophores, medusae, fish, etc. were removed and are not
included in our measurements.

Step 1. Wet weight determination

After washing with tap-water through a sieve to remove extraneous salts
(seawater, formalin and buffer), the sample is placed in a pre-calibrated
30 ml Gooch crucible with fritted glass disc. By adding slight pressure (ca.
5 psi) by means of a pressure pump to the wet Gooch crucible containing
the sample, interstitial water surrounding the plankton is removed with
relatively little detrimental effect on the organisms. There may be some loss
of internal salts to a varying degree. The sample is weighed immediately
after no more fluid is dripping and before any appreciable loss of weight
takes place due to evaporation.

Step 2. Displacement volume determination

Immediately following the wet weight determination, the precalibrated
Gooch crucible (and plankton) are immersed in a mercury bath. From the
amount of water titrated into the crucible, the displacement volume is
determined according to the method of Yentsch & Hebard (1957). The
results by this method give smaller but more accurate values than those

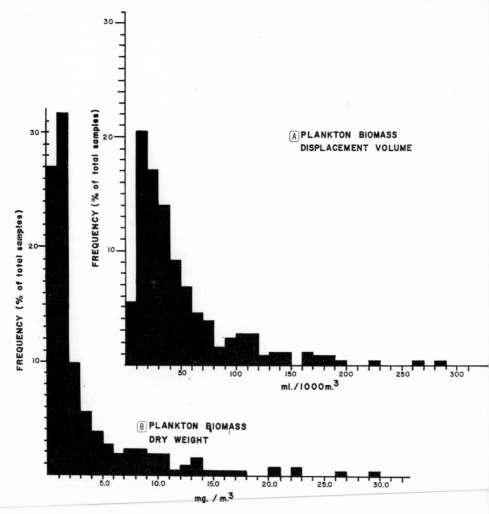

Figure 2a   Frequency distribution of plankton displacement volumes
grouped in 10 ml/1000 m³ intervals

Figure 2b   Frequency distribution of plankton wet weight grouped in
1.0 g/1000 m³ intervals

obtained by the conventional drip method which leaves much interstitial water clinging to the organisms.

Figure 2a is a frequency distribution of the total plankton samples according to size classes of displacement volumes. It shows that the majority of the samples yielded displacement volumes between 10 and 29.9 ml/1000 m^3. The frequency distribution is skewed towards the lower end and indicates that the higher the displacement volumes, the less frequently they occur. This may partly be attributed to our sampling coverage that is predominantly in warm regions.

Step 3. Dry weight determination

The dry weight is determined by heating the sample in the Gooch crucible at 60°C for one hour in a vacuum-oven at 15 psi. The sample is cooled to room temperature in a vacuum desiccator before weighing. The reader is referred to Lovegrove (1965) for a detailed discussion on procedures and the factors affecting the results.

Figure 2b is a frequency distribution of the total number of plankton samples arranged in size classes of dry weight. It is also heavily skewed towards the low range, since most of the dry weights are less than 2.0 mg/m^3.

Step 4. Ash weight determination

Ash weight is obtained by igniting the sample in a vacuum oven at 500°C for 2 hours to remove all organic matter, leaving only the inorganic residue of the organisms. We followed the procedure established by Sachs (1964). After cooling in a desiccator to room temperature, the sample is weighed and brushed onto a microscope slide to be examined and preserved for further study. More recently, we have employed Smith's technique (1967) in which a sodium hypochlorite solution is mixed with the dried sample to digest the organic matter before combustion.

Data Processing

The displacement volume, wet weight, dry weight, and ash-free dry weight values for each sample were transferred to punch cards along with various field data. An IBM Model 1800 computer was used to convert these raw sample parameters into absolute values (ml/1000 m^3 or mg/m^3) by taking into account the volume of water filtered and the sample aliquot used. The absolute data for each parameter, were then displayed on geographic plots and along selected cruise transects, using an IBM Model 1627 X-Y plotter.

GEOGRAPHIC VARIATIONS IN PLANKTON DISPLACEMENT VOLUMES

Despite the obvious shortcomings of volumetric comparisons for plankton abundance when abnormally large volumes are contributed by large organisms such as salps, medusae, siphonophores or fish, this method is still a commonly accepted measure of general plankton standing stock. Its greatest merit is the ease of measurement and nondestruction of the organisms.

Central waters

Geographic variations in the displacement volumes of North Atlantic zooplankton are shown in Figure 3. It is apparent that the central North Atlantic waters, approximately between 20°N and 40°N, is a plankton-poor region having displacement volumes of less than 25 ml/1000 m³. Lowest values are encountered in the southwestern part of the Sargasso Sea, where they are less than 10 ml/1000 m³. These oligotrophic waters extend almost from the western to the eastern margins of the central North Atlantic Ocean basin, except along the region off West Africa. The sparsity of plankton may be attributed to the depletion of the relatively small concentration of nutrients by phytoplankton and the lack of replenishment from deeper waters due to the hydrographic stratification during most of the year.

Winter overturning and nutrient enrichment of the euphotic zone result in a several-fold increase in plankton volume during spring. In a series of samples obtained during a January–February transect by CSS *Hudson* from the Grand Banks to the Caribbean Sea (Fig. 4a), we obtained volumes of 80 and 150 ml/1000 m³ at two stations north of the Gulf Stream. The low volumes (less than 25 ml/1000 m³) between latitudes 40°N and 33°N are typical for the northern Sargasso Sea. Somewhat higher volumes (30 to 80 ml/1000 m³) occur south of the east-west trending "oceanic front" located at about 30°N.

Clarke (1940) and Grice and Hart (1962) have also noted that seasonal fluctuations in the Sargasso Sea are relatively small compared to the slope and coastal waters off the northeastern United States. Clarke studied the zooplankton abundance at 9 stations between New York and Bermuda and concluded that (1) the ratios of displacement volumes between the coastal area, slope waters and Sargasso Sea was approximetaly 16 : 4 : 1; (2) the coastal plankton volumes in summer were 20 to 40 times greater than in winter; (3) the slope water plankton in summer were 10 times greater than

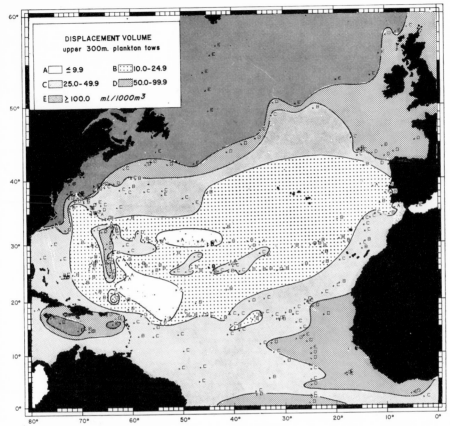

Figure 3 Geographical variations in zooplankton displacement volumes in the North Atlantic Ocean, based on plankton tows in the upper 300 m of water

in winter; and (4) there was no significant seasonal change in the Sargasso Sea. After averaging and weighting the results by ocean area and sampling depth, he obtained the following mean plankton volumes per 30-minute tow (using 75 cm scrim nets with 10 strands/cm): 194 ml for coastal water, 52 ml for slope water and 12 ml for the Sargasso Sea.

Jespersen (1924, Figs. 1 and 2) observed lowest displacement volumes in the southwestern Sargasso Sea (quadrant between 20°N and 30°N and 60°W and 70°W) for macroplankton in the upper 100 m. Jespersen (1935, Fig. 2) noted again in two Atlantic crossings that minimum quantities of macroplankton occured in the southwestern and central Sargasso Sea.

In the Bermuda region, Menzel and Ryther (1961) reported that the spring maxima in displacement volumes were 21, 6 and 4 times the minimum values, and Deevey (1962) found a 10-fold increase during the spring over

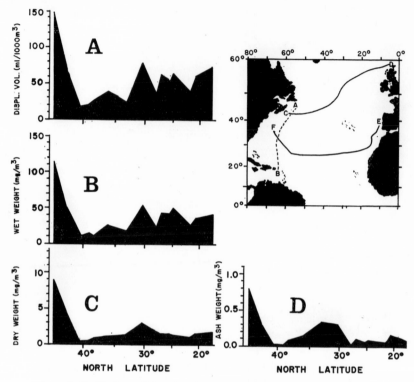

Figure 4 Latitude variation in plankton displacement volume (A), wet weight (B), dry weight (C), and ash weight (D) in western North Atlantic, January–February, 1968. CSS *Hudson* transect A–B on inset map

the lowest observed plankton volume. However, these maxima last only for short durations (March 25–April 21, 1958; March 25–April 7, 1959; March 8–April 7, 1960; March 1961; March 5–April 7, 1962), while for most of the year the average annual volumes remain below 25 ml/1000 m³. It is noteworthy that the plankton nets used by Menzel and Ryther (1961) and Deevey (1962) had a coarser mesh-aperture (366 μ) than ours (202 μ), but that their and our displacement volumes are of the same order of magnitude.

Cold-temperate and subarctic waters

A zone of generally high displacement volumes between 25 and 49.9 ml/ 1000 m³ can be found around the margins of the central North Atlantic waters. These are regions with strong boundary currents such as the Gulf Stream System, the North Atlantic Current, the Canaries Current, the North Equatorial Current System and the Antilles Current. In the Gulf Stream System, our plankton volumes ranged mostly from 25 to 49.9 ml/1000 m³ and they agree with the mean value of 30 ml/1000 m³ for four seasons obtained by Grice and Hart (1962).

Very high plankton volumes exceeding 100 ml/1000 m³ are common in slope and shelf waters and extend north of Cape Hatteras into the subarctic waters. Similar observations were made by Bigelow and Sears (1939), Clarke (1940), St. John (1958), Grice and Hart (1962) and others.

The highest plankton volumes in the North Atlantic are encountered between May and October in subarctic and cold-temperate waters. A maximum of 960 ml/1000 m³ occurred at 17°09′N and 35°50′W in June. Values greater than 50 ml/1000 m³ were generally found north of 45°N or roughly north of a broad arc from Cape Hatteras to the Bay of Biscay. These high densities frequently exceeding 100 ml/1000 m³ occurred along a transect of USCG *Yakutat* from New York to Scotland, June 11–25, 1962 (Fig. 5a). Other collections in subarctic waters during summer months also produced volumes higher than 100 ml/1000 m³.

Our observations are in general agreement with those of Kusmorskaya (1961), who reported high volumes between 100 and 500 ml/1000 m³ in the upper 200 m of water in a broad region between Newfoundland and the British Isles north of 40°N. Her samples were obtained with Juday nets (0.17 mm mesh-aperture) in spring and autumn of 1958.

The seasonal fluctuations of subarctic zooplankton in surface waters of the central Labrador Sea were studied by Kielhorn (1952) based on half-meter plankton nets (0.360 mm mesh-aperture) towed for 20 minutes at weekly intervals at Station Bravo (56°30′N, 51°00′W). If we assume that Kielhorn's semi-quantitative samples filtered roughly 100 m³ per haul, the displacements volumes range from 1000 to 7000 ml/1000 m³ from April to October, with peak volumes between May and July. The volumes were less than 200 ml/1000 m³ from November to March. Thus, Kielhorn's minimum volumes in winter correspond roughly with late summer volumes.

The results of Wiborg (1954), Østvedt (1955) and Hansen (1960) indicate that the seasonal variations in subarctic waters of the southern Norwegian

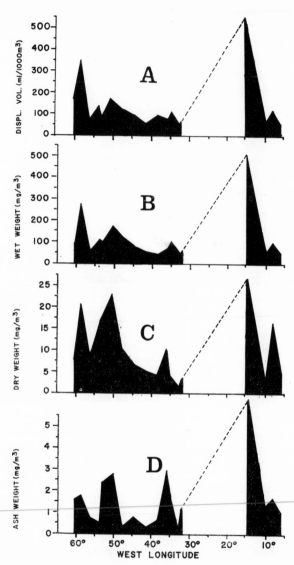

Figure 5 Variation in plankton displacement volume (A), wet weight (B), wet weight (C), and ash weight (D) in the mid latitudes of the North Atlantic June 11–25, 1962. USCG *Yakutat* transect C–D on inset map of figure 4

Sea are extremely large in magnitude—the summer maximum being thirty times greater than the winter biomass in the upper 50 m. This variation is due to a large extent to the massive seasonal vertical migration of the major zooplankton species, such as *Calanus finmarchicus* and *Pseudocalanus minutus*, that "hibernate" below 600 m.

Subtropical and tropical boudary currents

Moderately high displacement volumes over 50 ml/1000 m^3 are noted off West Africa extending almost halfway across the equatorial Atlantic between 20° and 5°N latitudes. This is the zone of divergence between the North Equatorial Current and the Equatorial Counter Current (Defant, 1936), where vertical mixing and nutrient enrichment of the epipelagic waters contribute to a rich plankton biomass. Kanaeva (1965) reported moderate biomass for May–June 1961 and October–November 1962 and spring blooms for April–May 1969 for those regions (Fig. 7). Jespersen (1924, 1935) also reported high densities in these waters.

The *Yakutat* transect from Spain to the Canary Islands to Bermuda (Fig. 6a) represents the boundary conditions between the eutrophic North Equatorial Current and the oligotrophic Sargasso Sea. Our values along this transect range from 14.3 to 57.5 ml/1000 m^3.

A number of investigators (Menzel and Ryther, 1961; Tranter, 1961; Wickstead, 1963; Hopkins, in press) have noted that similar values of displacement volumes (and wet weight) are obtained regardless of the mesh size and mouthopening of the plankton net used. In comparative field studies using paired plankton samplers with mesh-apertures varying from 76 microns to 366 microns, they concluded independently that the resulting values of displacement volumes (or wet weight) for each set of hauls fell within remarkably close ranges. It appears that although a fine-meshed net will retain a larger number of smaller organisms, a coarse-meshed net will yield a roughly equivalent biomass by catching the larger individuals with greater efficiency.

Our present distribution patterns of plankton volumes in the North Atlantic agree in a broad sense with the regional studies of Jespersen (1924, 1935), Hentschel (1942), Friedrich (1950) and Kusmorskaya (1961). Table 2 provides comparative values of displacement volumes and wet weights obtained by various investigators for three broad provinces in the North Atlantic. It is evident that the regional variations in zooplankton abundance

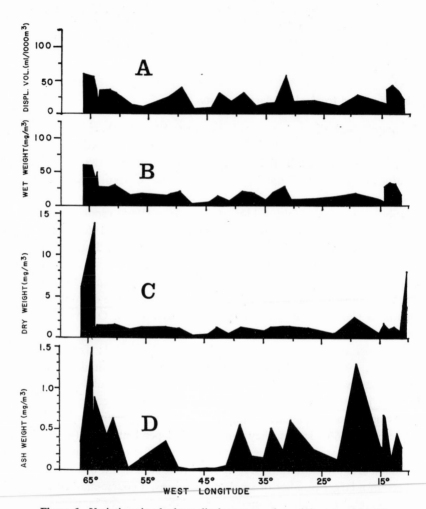

Figure 6   Variations in plankton displacement volume (A), wet weight (B), dry weight (C), and ash weight (D) in the low latitudes of the North Atlantic July 19–August 9, 1962. USCG *Yakutat* transect E–F on inset map of figure 4

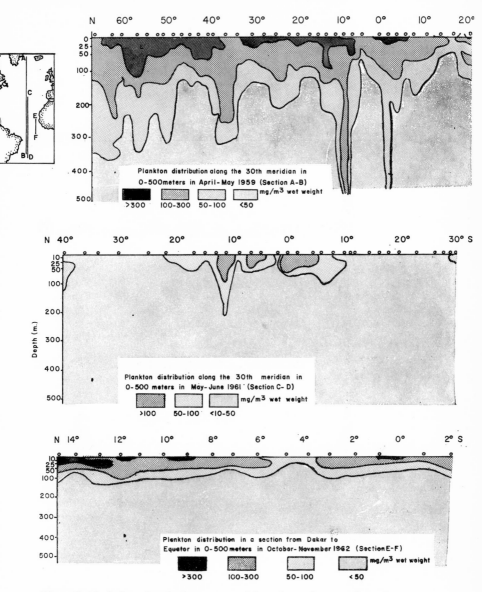

Figure 7 Variations in plankton wet weights along three north-south transects. (From: I. P. Kanaeva, 'On the quantitative distribution of plankton in the Atlantic Ocean', *Trudy VNIRO*, **57,** pp. 333–343, 1965)

Table 2 Comparison of displacement volumes and wet weights in various regions of the North Atlantic

(1 = summer; 2 = fall; 3 = winter; 4 = spring; 5 = yearly mean)

I *Boreal North Atlantic and Neritic Waters*

Region	Displ. vol. ml/1000 m^3	wet wt. mg/m^3	Net mesh aperture (mm)	Depth range (m)	Reference
Iceland Coast					Jespersen (1940)
(1)	250		stramin	0–50	Idem
(4)	450		stramin	0–25	Idem
Southern Norwegian Sea					
(1)	340		0.366	0–100	Wiborg (1954)
(3)	10		0.366	0–100	Idem
(4)	230		0.366	0–100	Idem
Labrador Current					
(2) 43° N		1072	0.17	0–200	Yashnov (1961)
Norwegian Sea (5)		>500	0.17	0–100	Kanaeva (1965)
Cold-temperate Subarctic waters (5)	>100	>100	0.202	0–300	This paper
Western North Atlantic					
Coastal	8100		0.158	0–25	Riley (1939)
Slope water (4)	4300		0.158	0–50	Idem
Slope water (1)		430–1600	0.158	0–400	Riley & Gorgy (1948)
Coastal (5)	540		10 strands/cm	0–85	Clarke (1940)
Offshore (5)	400				Idem
Gulf of Maine (3)	120		Front: 29–38 meshes/inch	variable	Bigelow & Sears (1939)
(1)	260		Rear: 48–54 meshes/inch	variable	Idem
Cape Cod– Chesapeake Bay coastal shelf					
(1)	700–800			variable	Idem
(3)	400			variable	Idem

Table 2 (*cont.*)

Region	Displ. vol. ml/1000 m³	Wet wt. mg/m³	Net mesh aperture (mm)	Depth Range (m)	Reference
Georges Bank					
(1)	1500		0.366	0–25	Clarke & Bishop (1948)
(3)	200		0.366	0–25	Idem
Cape Hatteras-Cape Fear					
(1)	280		0.360	Variable over shelf	St. John (1958)
Continental Slope 38°–41° N					
(2)	328		0.170	0–200	Yashnov (1961)
New York-Bermuda Coastal water (5)	1070		0.230	0–200 or less	Grice & Hart (1962)
Slope water (5)	270		0.230	0–200	Idem

II   *Central waters (Sargasso Sea)*

Region	Displ. vol. ml/1000 m³	Wet wt. mg/m³	Net mesh aperture (mm)	Depth Range (m)	Reference
Sargasso Sea					
(5)	20		0.230	0–200	Grice & Hart (1962)
Sargasso Sea Bermuda					
(5)	28		0.203–0.366	0–500	Menzel & Ryther (1961)
S. W. Sargasso Sea 20°–37° N (2)		156.7	0.17	0–200	Yashnov (1961)
Sargasso Sea					
(1)		45	0.158	0–400	Riley & Gorgy (1948)
Sargasso Sea					
(5)		50–100	0.170	0–100	Kanaeva (1965)
Sargasso Sea					
(5)		<50–100	?	0–300	Hela & Laevastu (1961)
Sargasso Sea					
(5)	10–25	<10–25	0.202	0–300	This paper

Table 2 *(cont.)*

III *Boundary Currents*

Region	Displ. vol. ml/1000 m³	Wet wt. mg/m³	net mesh aperture (mm)	Depth Range (m)	Reference
Florida Strait (4)	20		0.158	0–150	Riley (1939)
Florida Strait (4)	20		0.158	150–300	Idem
Gulf Stream off Florida (4)	50		0.158	0–100	Idem
Gulf Stream off Georgia (4)	70		0.158	0–150	Idem
Gulf Stream		137	0.158	0–400	Riley & Gorgy (1948)
Gulf Stream		250–500	0.170	0–100	Kanaeva (1965)
Gulf Stream betw. N. Y. – Bermuda	30		0.230	0–200	Grice & Hart (1962)
Gulf Stream (north) 40°–43° N		143	0.17	0–200	Yashnov (1961)
Gulf Stream (south) 35°–37° N		114	0.17	0–200	Idem
N. Equat. Current 16°–19° N		201	0.17	0–200	Idem
Canary Current 27°–36° N		161	0.17	0–200	Idem
Equat. Current (1)	100–>300		0.336	0–~50	Mahnken (1969)
Equat. Current region (5)		100–500	0.17	0–100	Kanaeva (1965)
Equat. Current region (5)		100–300		0–300	Hela & Laevastu (1961)
All boundary Currents around Sargasso Sea (5)	25–100	25–100	0.202	0–300	This paper

are of sufficiently large magnitude that they mask variations due to diurnal vertical migration, local patchiness, nonsynoptic sampling and sampling gear types. Thus, it is possible to make the following general estimates of relative zooplankton abundances: displacement volumes over $200\,ml/1000\,m^3$ (or over $200\,mg/m^3$ wet weight) in boreal and neritic waters of the North Atlantic; 50 to $200\,ml/1000\,m^3$ in the boundary currents; and less than $50\,ml/1000\,m^3$ in the central waters.

GEOGRAPHIC VARIATIONS IN PLANKTON WET WEIGHTS

Wet weight as a biomass measure is the method used by many Soviet planktologists (e.g. Bogorov, 1967; Kanaeva, 1965; Vladimirskaya, 1966) and Riley and Gorgy (1948), Hela and Laevastu (1961), McAllister (1961), Tranter (1962) and others. A given wet weight should yield a slightly higher numerical value than its equivalent displacement volume, because the protoplasmic weight of zooplankton is heavier than sea water. In practice we have found that the reverse is true, as displacement volumes are consistently higher than their wet weight. Ahlstrom and Thrailkill (1963, p. 68) have also made a similar observation for copepods in their following results:

Determinations made on 2,000 *Calanus helgolandicus*:

Wet displacement volume	2.8 ml
Interstitial liquid	1.5 ml
Copepods only	1.3 ml
Wet wight	2.68 g
Blotted weight	1.15 g
Dry weight	0.1694 g
Ash weight	0.0077 g
Percent ash-dry weight	4.55%

The reason for the above-mentioned discrepancy is not yet clear; perhaps interstitial air may have been introduced during our wet weight determinations resulting in larger displacement volumes.

Figure 8 is a map of wet weights expressed as percentages of their displacement volumes. It is noteworthy that the wet weights approximate their displacement volumes most closely in the subarctic and north-equatorial waters where relatively high biomass is prevalent. The largest difference in weight/volume occurs in the oligotrophic central waters, where the plankton tends to be highly diverse and small. We have no conclusive explanation for

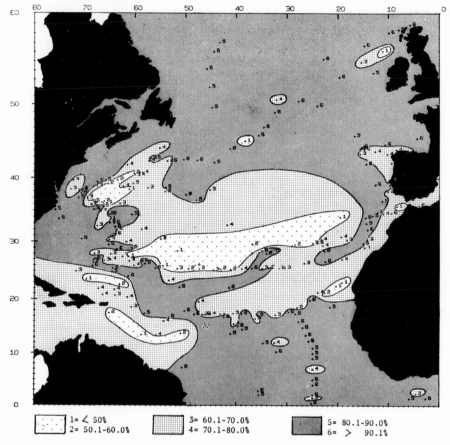

Figure 8 Distribution of wet weights in terms of percentages of their displacement volumes. Note that higher percentages occur in subartic and north-equatorial regions

these observations. A tentative theory is that the eutrophic regions are composed of larger proportions of crustaceans or shellbearing plankters that are denser than "gelatinous" organisms.

Central waters

In general, the distributional patterns of wet weights in Figure 9 are very similar to those for displacement volume. They agree very closely to the North Atlantic estimates by Cushing Corlett (see Walford, 1958, Fig. 15).

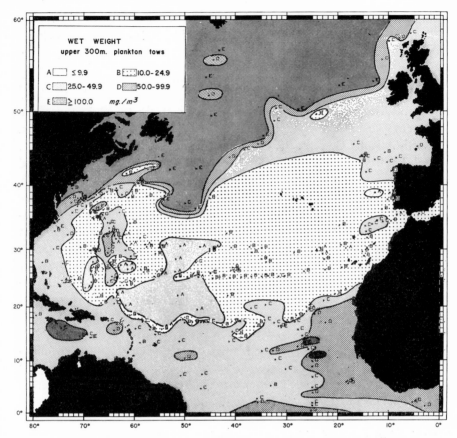

Figure 9 Geographical variations in zooplankton wet weight based on plankton tows in the upper 300 m of water

Low plankton wet weights of less than 25 mg/m³ occur in the central North Atlantic between 20° and 40°N Latitudes. The boundaries of the oligotrophic region are similar to those shown for displacement volumes in Figure 3. Minimum weights appear again in the southwestern Sargasso Sea.

In the western North Atlantic, the Hudson north–south transect of January–February 1968 (Fig. 4b) indicates somewhat higher wet weights (25–100 mg/m³) than during other months of the year.

Hela and Laevastu's (1961) and Kanaeva's distributional patterns for the central waters are basically in agreement with ours although their values tend to be higher. For the central waters, our observed values of less than 25 mg/m³ compare with less than 100 mg/m³ of Hela and Laevastu (1961) and with 50–250 mg/m³ of Kanaeva (1965).

Cold-temperate and subarctic waters

Intermediate wet weights (25.0 to 49.9 mg/m³) are found in the active current systems surrounding the central waters, as in the Gulf Stream system, North Atlantic Current, the Canaries Current, North Equatorial Current and Antilles Current. In the slope and shelf waters off New England and the maritime provinces of Canada even higher standing stocks occur, especially during the warmer months. Wet weights greater than 50 mg/m³ are frequently encountered in the cold-temperate waters from Cape Cod northwards and in subarctic waters between Newfoundland and the British Isles, as exemplified in the *Yakutat* transect from New York to Scotland (Fig. 5 b).

Kanaeva (1965) and Hela and Laevastu (1961) again observed higher biomasses of 250–500 mg/m³ and 200–400 mg/m³, respectively, for the subarctic North Atlantic region. Kanaeva showed that during April and May the plankton biomass in subarctic waters amounted to 300–700 mg/m³ or more in the upper 100 m (Fig. 7).

Vladimirskaya's (1966) seasonal study of zooplankton in the Newfoundland area indicated a broad zone where high biomass of 200–300 mg/m³ in the upper 100 m extended north-eastward from the Grand Banks, whereas such abundances were limited to the Grand Banks during autumn.

Subtropical and tropical boundary currents

A less extensive, but equally rich region (50–99.9 mg/m³) occurs in the divergence zone between the North Equatorial Current and the Equatorial Countercurrent. Wet weights of over 100 mg/m³ may be encountered in the western portion and they are probably associated with the current upwelling off West Africa.

The *Yakutat* transect from Spain to the Canary Islands to Bermuda (Fig. 6 b) yielded rather low standing stocks, mostly below 25 mg/m³, except for somewhat higher values off southern Spain and near Bermuda.

Kanaeva (1965) and Hela and Laevastu (1961) reported biomass weights of 250–500 mg/m³ and 200–400 mg/m³ in the corresponding eutrophic regions off West Africa. These values are on the average several times higher than our observations.

Riley and Gorgy (1948) determined the following average wet weights of zooplankton in the upper 400 m of water (or surface to near bottom in coastal water): 840 mg/m³ in the coastal area south of Cape Cod, 150 mg/m³ in the Gulf Stream and 45 mg/m³ in the Sargasso Sea. Their samples were based on Clarke–Bumpus samplers (with 158 μ mesh-aperture) towed at 2 knots for 30 minutes.

GEOGRAPHIC VARIATIONS IN PLANKTON DRY WEIGHTS

"Dry weight" is considered by many planktologists (Krey, 1958; Corlett, 1957, 1961; Menzel and Ryther, 1961; Hopkins, 1969a, b, and others) to be a more reliable measure of organic biomass and plankton abundance than "displacement volume" or "wet weight". On the other hand, the method is destructive and, therefore, it is necessary either to study the samples first or to collect replicate samples for taxonomy.

The dry weights in the North Atlantic (Fig. 10) appear to be of lesser geographic variability than those shown for the other three biomass measures, although this may perhaps be an artifact of the selection of contour intervals. Most of the low- and mid-latitude regions have dry weights between 0.1 and 3.9 mg/m³. These values are found not only in the central waters, but also in the boundary currents of the Antilles Current, sections of the Gulf Stream system and North Atlantic Current, Canaries Current and the western part of the North Equatorial Current (Fig. 4c and 6c).

Higher dry weights are encountered in the cold-temperate and subarctic waters from the slope and shelf waters north of Cape Hatteras to the British Isles. In these northern areas the biomass varies mostly from 4.0 to 19.9 mg/m³ dry weight (Fig. 5c).

Another region of relatively high dry weights occurs in the North Equatorial Current off West Africa, where the values also range from 4.0 to over 20 mg/m³. A maximum of 52.6 mg/m³ was found in the equatorial Atlantic (00°28'N, 25°14'W).

We consider the high dry weights in the high-latitude and equatorial areas to be partly due to considerable densities of shell-bearing euthecosomatous pteropods, foraminifera (and sometimes diatoms) and the exoskeletons of crustaceans.

There are very few comparable regional surveys in other oceans. Bogorov (MS) reported the results of Soviet expeditions on the geographic variations

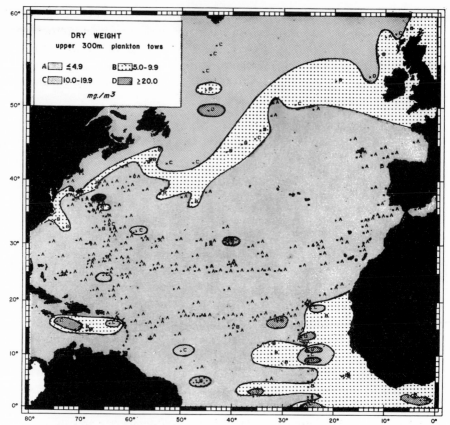

Figure 10 Geographical in zooplankton dry weight based on plankton tows in the upper 300 m of water

in plankton dry weights for the upper 100 m in the Indian Ocean, based on sampling with Juday nets (0.17 mm mesh-aperture). The biomass weights were less than 2.5 mg/m³ in the central waters between Madagascar and Australia and south of 20°S. Intermediate values of 5 to 10 mg/m³ were found in the subtropical convergence belt between 15°S and the equator. High biomass of 10 to 15 mg/m³ or higher was prevalent in regions of coastal upwelling or current divergence systems.

Menzel and Ryther (1961) and Deevey (1962) reported in their seasonal study of plankton in the Sargasso Sea off Bermuda that the annual mean dry weight amounted to 2.16 mg/m³ (= 1.08 g/m²), while the short-lived spring maxima reached 16.0 mg/m³ in April 21, 1957; 3.9 mg/m³ in March 25,

1959; 2.6 mg/m^3 in April 7, 1960; 5.2 mg/m^3 on March 16, 1961; 4.9 mg/m^3 on April 7, 1962. These observations are in general agreement with our results.

Our dry weight values for the North Atlantic are on the average much higher than those observed by Hopkins (1969a) for the Arctic Ocean. He has estimated mean dry weight concentrations of 0.62 mg/m^3 for the Arctic surface layer (0–200 m), 0.14 mg/m^3 for the Atlantic layer (200–900 m), and 0.04 mg/m^3 for the Arctic deep-water mass (>900 m). Hopkins' (1969b) average value in the upper 500 m of the East Greenland Current of 9.04 mg/m^3 is still lower than our dry weights in subarctic waters.

GEOGRAPHIC VARIATIONS IN PLANKTON ASH-FREE DRY WEIGHTS

Ash-free dry weight is obtained by subtracting "ash weight" from "dry weight". Thus it is more accurate than the previous three methods as a measure of the amount of organic matter, since it does not include the inorganic fraction contributed by skeletal material of foraminifera, radiolaria, pteropoda, etc.

As in the previously discussed biomass parameters, the sparsely populated central waters yield the lowest ash-free dry weights of less than 1 mg/m^3 (Fig. 11). The boundary currents encircling the Sargasso Sea are somewhat richer (1.0–2.5 mg/m^3). During the *Hudson* winter transect comparatively high ash-free dry weights were observed and they account for the irregular distributional patterns in the western Sargasso Sea shown in Figure 11. Menzel and Ryther reported maximum ash-free dry weights of 11.7, 3.5 and 2.2 mg/m^3 during the respective 1958, 1959 and 1960 spring zooplankton bursts in the waters off Bermuda, while the year-round averages are generally lower than 1.5 mg/m^3.

Riley and Gorgy (1948, Fig. 18) observed that the organic content of summer zooplankton in the north-central Sargasso Sea ranged between 1.1 and 2.2 g/m^2 (= 2.2 and 4.4 mg/m^3).

The highest amounts of organic matter are encountered again in the subarctic and cold-temperate regions from Cape Hatteras to the Grand Banks to north of Scotland. Another region of equivalent plankton standing stock is the eastern portion of the equatorial waters off West Africa. In both of these eutrophic regions, the ash-free dry weights commonly range between 2.5 and 10 mg/m^3. Values higher than 10 mg/m^3 are noted at a number of stations, but their geographic extent can not be mapped with certainty.

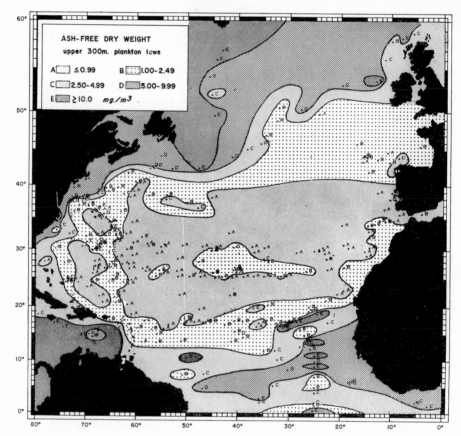

Figure 11 Geographic variations in zooplankton ash-free dry weight
based on plankton tows in the upper 300 m of water

GEOGRAPHIC VARIATIONS IN PLANKTON ASH WEIGHTS

The distributional patterns of ash weights shown in Figure 12 again reflect
the relative densities of plankton in the North Atlantic. The ash weights
are predominantly composed of the inorganic remains of skeletal material
from foraminifera, pteropoda, heteropoda, radiolaria, diatoms, etc. Values
greater than 1 mg/m³ appear in the subarctic waters, where the shells of the
pteropods *Limacina retroversa* and *L. helicina* are large components. Comparatively high values occur off West Africa and the north equatorial waters,
where both foraminiferal and pteropod shells contribute significantly to

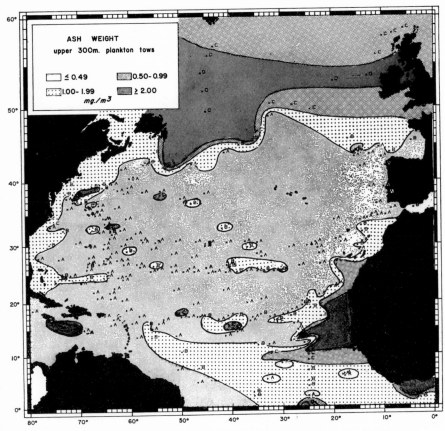

Figure 12 Geographic variations in zooplankton ash weight based on plankton tows in the upper 300 m of water

the ash wights. Over the greater part of the North Atlantic, the ash weights are generally lower than 0.5 mg/m³, although somewhat higher values (0.5–1.0 mg/m³ are encountered in slope and shelf waters off New England and Nova Scotia and in some stations in the central Sargasso Sea.

BIOMASS RATIOS

We have calculated the ratios between the mean values of the displacement volumes, wet weights, dry weights and ash-free dry weights for our samples and compared them with ratios obtained by other investigators (Table 3).

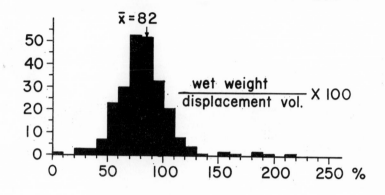

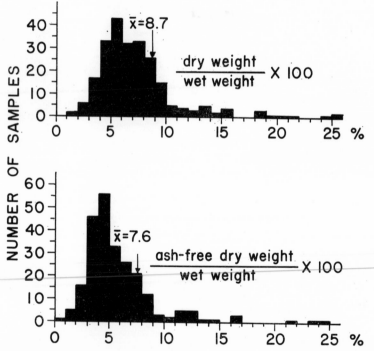

Figure 13 Frequency distributions of one biomass parameter as a percentage of another parameter

Table 3   Biomass Ratios according to various investigators

Displ. vol.	Wet wt.	Dry wt.	Ash-free dry wt.	
15.9	13.0	1.1	1	This paper
10	10	1	—	Hopkins (in press)
—	—	1.25	1	Beers (1966)
26.5	—	1	—	Bsharah (1959)
18	—	1.4	1	Menzel and Ryther (1961)
				Beers (1966):
—	7.4	1	—	Copepods
—	6.3	1	—	Euphausids-mysids
—	5.5	1	—	Other crustacea
—	14.7	1	—	Chaetognaths
—	7.0	1	—	Fish/Fish larvae
—	10.3	1	—	Polychaetes
—	25.0	1	—	Siphonophores
—	23.0	1	—	Hydromedusae
—	3.9	1	—	Pteropods

Figure 13 shows three frequency distributions of one biomass parameter as a percentage of another parameter, and their significance is as follows:

1. The average of 246 wet weights amounted to 82% of the displacement volumes.

2. The average of 246 dry weights amounted to 8.7% of the wet weights.

3. The average of 246 ash-free dry weights amounted to 7.6% of the wet weights.

4. Thus, the mean ratios between displacement volumes, wet weight, dry weights and ash-free dry weights for 246 samples are 15.9 : 13.0 : 1.1 : 1.

These values are merely intended as gross approximations of the relative magnitudes of our biomass parameters. They do not necessarily prevail from region to region or from season to season, but are useful first-order-estimates of plankton abundance ratios.

GROUP DIVERSITY

The following 27 taxonomic groups were recorded and enumerated from each of 176 plankton samples counted and they are listed in decreasing frequency of occurrence:

	Number of occurrences
Copepoda	176
Ostracoda	171
Chaetognatha	169
Foraminifera	156
Siphonophora	154
Pteropoda	153
"Other Protozoa"	150
Fish eggs	149
Larvacea	147
Medusae	133
Amphipoda	133
Polychaeta	128
Euphausiacea	127
Mysidacea	124
Thaliacea	99
Fish larvae	79
Cyphonautes	59
Decapoda	56
Heteropoda	43
Gastropoda larvae	34
Bipinnaria	32
Cladocera	31
Cephalopoda (juvenile)	17
Brachyuran larvae	14
Ctenophora	10
Phyllosoma	7
Stomatopoda	1

All major groups were found in all regions of the North Atlantic, indicating that geographic discontinuity occurs predominantly at the species or subspecies level. Yet, we do note that the total number of plankton groups recorded in each sample is limited and seldom exceeded 22. A plot of total group diversity showed some interesting distributional patterns (Fig. 14).

The highest total diversity of 16 groups or more are found in the central waters and to some extent in equatorial waters. Intermediate values (13 to 15 groups) occur in the boundary currents surrounding the central waters. Lowest total diversity (12 or less groups) are encountered in cold-temperate and subarctic waters.

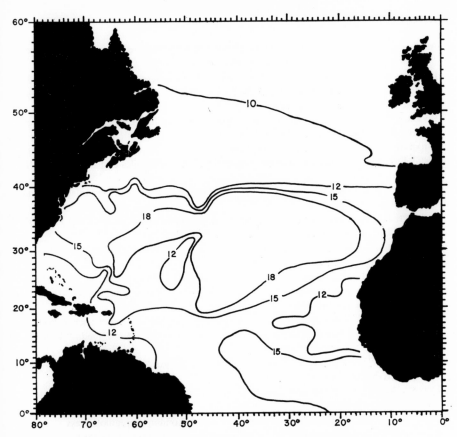

Figure 14   Diversity of zooplankton groups (out of a total of 27 taxonomic groups) in the North Atlantic Ocean

Acknowledgements

We are grateful to Professor Maurice Ewing and other colleagues, too numerous to mention by name, at Lamont–Doherty Geological Observatory for their aid in the field and laboratory; to the officers and men of the U. S. Coast Guard who participated in the plankton sampling program on USCGC *Rockaway* and *Yakutat*; to Dr. Grant Bartlett and Gustavs Vilks of the Bedford Institute of Oceanography for enabling us to participate in a cruise on CSS *Hudson*. Mr. Stanley Harrison has provided invaluable help in our sampling program. This study received support from the Office of Naval Research, Contract 00014-67-0108-0004.

48 *Fertility of the Sea*

References

AHLSTROM, E. H., and THRAILKILL, J. R. (1962). Plankton volume loss with time of pre-servation. *Calif. Coop. Oceanic Invest. Rpts.*, **9**, 57–73.

BIGELOW, H. B., and SEARS, M. (1939). Studies of the waters of the continental shelf, Cape Cod to Chesapeake Bay. III. A volumetric study of the zooplankton. *Mus. Comp. Zool. Harvard Mem.*, **54**, 4, 189–373.

BOGOROV, V. G. (1967). Biological transformation and exchange of energy and matter in the ocean. *Okeanologiya*, **7**, 5, 649–665 (In Russian).

BSHARAH, L. (1957). Plankton of the Florida Current. V. Environmental conditions, standing crop, seasonal and diurnal changes at a station 40 miles east of Miami. *Bull. Mar. Sci. Gulf and Caribb.*, **7**, 3, 201–251.

CLARKE, G. L. (1940). Comparative richness of zooplankton in coastal and offshore areas of the Atlantic. *Biol. Bull.*, **78**, 226–255.

CORLETT, J. (1961). Zooplankton observations at Ocean Weather Stations I and J in 1957 and 1958. *Cons. Internat. Explor. Mer., Rapp. Proc. Verb.*, **149**, 200–201.

CUSHING, D. H. (1959). The seasonal variation in oceanic production as a problem in population dynamics. *J. Conseil.*, **24**, 455–464.

DEEVEY, G. B. (1962). The annual cycle in quantity and composition of the zooplankton in the Sargasso Sea, March 1961 to April 1962. *AEC Rept. Contr.* AT (30-1) 2646, Bermuda Biological Station.

DEFANT, A. (1936a). Ausbreitungs- und Vermischungsvorgänge im antarktischen Boden-strom und in Subantarktischen Zwischenwasser. *Deutsche Atl. Exped. "Meteor"* 1925–27 *Wiss. Erg.*, **6**, 2, 55–96.

DEFANT, A. (1936b). Die Troposphära. *Deutsche Atl. Exped. "Meteor"* (1925–27, *Wiss. Erg.*, **6**, 1, 3, 289–411.

DEFANT, A. (1938). Aufbau und Zirkulation des Atlantischen Ozeans. *Sitz. Ber. Preuss. Akad. Wiss., Phys.-Math. Kl.*, **15**, 29 pp.

DEFANT, A. (1961). *Physical Oceanography*, v. 1 and 2. London: Pergamon Press.

FOXTON, P. (1956). The distribution of the standing crop of zooplankton in the southern ocean. *Discovery Rpts.*, **28**, 191–236.

FRASER, J. H. (1968). Standardization of zooplankton sampling methods at sea. *UNESCO, Monograph on oceanographic methodology* no. 2, 147–174.

FRIEDRICH, H. (1950). Versuch einer Darstellung der relativen Besiedlungsdichte in den Oberflächenschichten des Atlantischen Ozeans. *Kieler Meeresforsch.*, **7**, 2, 108–121.

GRICE, G. D., and HART, A. D. (1962). The abundance, seasonal occurrence and distri-bution of the epizooplankton between New York and Bermuda. *Ecol. Monogr.*, **32**, 287–309.

HANSEN, V. K. (1960). Investigations on the quantitative and qualitative distribution of zooplankton in the southern part of the Norwegian Sea. *Medd. Danmarks Fisk. Havunder.*, Ny serie, **2**, 23, 3–53.

HELA, I., and LAEVASTU, T. (1961). *Fisheries Hydrography*. Fishing News (Books) Ltd., London, 5–137.

HENTSCHEL, E. (1942). Eine biologische Karte des Atlantischen Ozeans. *Zool. Anzeiger*, **137**, 7/8, 103–123.

HOPKINS, T. L. (1969a). Zooplankton standing crop in the Arctic Basin. *Limnol. Oceanogr.*, **14**, 1, 80–85.

HOPKINS, T. L. (1969a). Zooplankton biomass related to hydrography along the drift track of Arlis II in the Arctic Basin and the East Greenland Current. *J. Fish. Res. Bd. Canada*, **26**, 305–310.

HOPKINS, T. L. (in press). Zooplankton standing crop in the Pacific sector of the Antarctic.

JESPERSEN, P. (1924). On the quantity of macroplankton in the Mediterranean and the Atlantic. *Internat. Rev. Hydrobiol. Hydrogr.*, **12**, 1/2, 102–115.

JESPERSEN, P. (1935). Quantitative investigations on the distribution of macroplankton in different oceanic regions. *Dana Rpt.* 7, 3–44.

JESPERSEN, P. (1940). Investigations on the quantity and distribution of zooplankton in Icelandic waters. *Medd. Danmarks Fisk. Havunder.*, Serie: Plankton, 3, 5, 3–76.

JESPERSEN, P. (1954). On the quantities of macroplankton in the North Atlantic. *Medd. Danmarks Fisk. Havunder.* Ny Serie, **1**, 2, 3–12.

KANAEVA, I. P. (1965). On the quantitative distribution of plankton in the Atlantic Ocean. *All-Union Sci. Res. Inst. Pisc., VNIRO Trans.*, **57**, 333–343 (In Russian).

KIELHORN, W. V. (1952). The biology of the surface zone zooplankton of a Boreo-Arctic Atlantic Ocean area. *J. Fish. Res. Bd. Canada*, **9**, 5, 223–264.

KREY, J. (1958). Chemical determinations of net plankton, with special reference to equivalent albumen content. *J. Mar. Res.* **17**, 312–324.

KUSMORSKAYA, A. P. (1961). Distribution of plankton in the North Atlantic in spring and autumn 1958. *Cons. Internat. Explor. Mer. Rapp. Proc., Verb.*, **149**, 183–187.

LOVEGROVE, T. (1966). The determination of the dry weight of plankton and the effect of various factors on the values obtained. In: Barnes, H., (edit.) *Some contemporary Studies in Marine Science*, George Allen and Unwin Ltd., London, 429–467.

MCALLISTER, C. D. (1961). Zooplankton studies at ocean weather station "P" in the Northeast Pacific Ocean. *J. Fish. Res. Bd. Canada*, **18**, 1, 1–29.

MCEWEN, G. F., JOHNSON, M. W. and FOLSOM, T. R. (1954). A statistical analysis of the performance of the Folsom plankton sample splitter, based upon test observations. *Arch. Meteorol. Geophys. u. Bioklimatol., ser. A, Meteorol. u. Geophys.*, 7, 502–527.

MENZEL, D. W. and RYTHER, J. H.(1961). Zooplankton in the Sargasso Sea off Bermuda and its relation to organic production. *J. Conseil*, **26**, 3, 250–258.

ØSTVEDT, O. J.(1955). Zooplankton investigations from weather ship M in the Norwegian Sea, 1948–49. *Hvalråd. Skr.*, **40**, 5–93.

REID, J. L., Jr. (1962). On circulation, phosphate-phosphorus content, and zooplankton volumes in the upper part of the Pacific Ocean. *Limnol. Oceanogr.*, 7, 3, 287–306.

RILEY, G. A. and GORGY (1948). Quantitative studies of summer plankton populations of the Western North Atlantic. *J. Mar. Res.*, 7, 100–121.

RYTHER, J. H. (1969). Photosynthesis and fish production in the sea. *Science*, **166**, 3901, 72–76.

SACHS, K. N., CIFELLI, R. and BOWEN, V. T. (1964). Ignition to concentrate shelled organisms in plankton samples. *Deep Sea Res.*, **11**, 621–622.

SMITH, R. K. (1967). Ignition and filter methods of concentrating shelled organisms. *J. Paleont.*, **41**, 1288–1291.

St. John, P. A. (1958). A volumetric study of zooplankton distribution in the Cape Hatteras area. *Limnol. Oceanogr.*, 3, 4, 387–397.

Tranter, D. J. (1962). Zooplankton abundance in Australasian waters. *Australian J. Mar. and Freshwater Res.*, **13**, 2, 106–142.

Vladimirskaya, Y. V. (1966). Quantitative distribution and seasonal dynamics of zooplankton in the Newfoundland area. *Okeanol. Issled.*, **13**, 137–142. (In Russian.)

Walford, L. A. (1958). *Living resources of the sea.* Ronald Press, New York (321 pp.).

Wiborg, K. F. (1954). Investigations on zooplankton in coastal and offshore waters off western and northwestern Norway. *Fiskeridir. Skr. Havundersøk.*, **1**, 1, 7–246.

Wiborg, K. F. (1955). Zooplankton in relation to hydrography in the Norwegian Sea. *Fiskeridir. Skr. Havundersøk.*, **11**, 4, 5–66.

Winsor, C. P. and Clarke, G. L. (1940). A statistical study of variation in the catch of plankton nets. *J. Mar. Res.*, 3, 1, 22–24.

Yentsch, C. S. and Hebard, J. F. (1957). A gauge for determining plankton volume by the mercury immersion method. *J. Conseil*, **22**, 184–190.

Additional References

Clarke, G. L. and Bishop, D. W. (1948). The nutritional value of marine zooplankton with a consideration of its use as an emergency food. *Ecology*, **29**, 1, 54–71.

Mahnken, C. V. (1969). Primary organic production and standing stock of zooplankton in the tropical Atlantic Ocean-Equalant I and II. *Bull. Mar. Sci.*, **19**, 3, 550–567.

Riley, G. A. (1939). Plankton studies II. The western North Atlantic May–June, 1939. *Jour. Mar. Res.*, **2**, 2, 145–162.

Wickstead, J. H. (1963). Estimates of total zooplankton in the Zanzibar area of the Indian Ocean with a comparison of the results with two different nets. *Proc. Zool. Soc. London*, **141**, 577–608.

Yaschnov, V. A. (1961). Vertical distribution of the mass of zooplankton in the tropical region of the Atlantic Ocean. *Dok. Akad. Nauk. S.S.S.R.*, **136**, 3, 705–708.

Seasonal variation in the physico-chemical composition of sea water in Paradise Harbor—West Antarctica

NORBERTO L. BIENATI and RUFINO A. COMES

Instituto Antártico Argentino
Scientific Department
Chemical Section

Abstract

The present paper deals with oceanographic observations carried out in Paradise Harbor, an area adjoining the Southern Ocean, where the Scientific Station Almirante Brown is located. This is based on surface observations made during 1965 and 1967, in sea water areas, and down to depths of 90 meters in 1968. The samples collected were analysed to determine salinity, dissolved oxygen, pH, total alkalinity and nutrients (phosphates and nitrites), also measurements on temperature (three daily observations) at different layers such as 5, 10, 15, 20, 25, 50, 75 and 90 meters were carried out. Comparative tables and diagrams were plotted with all the data available showing the individual variation of the components analysed, sometimes associated in between.

The first part of the research work is presented here, and consists of a broad report on observational data without either giving full details of the phenomena or entering into the speculative phase of the work accomplished. A second part is being prepared, base on more observations and detailed analysis undertaken in the years 1969 and 1970.

Resumen

El trabajo se refiere a las observaciones oceanográficas desarrolladas en la bahía de Puerto Paraíso, área dependiente del Océano Sur, donde se encuentra instalada la Estación Científica Almirante Brown.

Las observaciones son de superficie en los años 1965 y 1967; y en aguas hasta 90 metros de profundidad en el año 1968. Las muestras fueron sometidas a análisis de salinidad, oxígeno disuelto, pH, alcalinidad total y nutrientes (fosfatos y nitritos), así como también las mediciones de temperatura (3 observaciones diarias) y temperature de los distintos estratos de 5, 10, 15, 20, 50, 75 y 90 metros de profundidad, estos últimos registrados durante los diferentes muestreos.

Con todos los datos obtenidos se confeccionaron distintos tipos de gráficos comparativos e interpretativos, que reflejan la variación individual, o a veces asociada, de los elementos analizados. La primera parte del trabajo, que es la que aquí se presenta, es un reporte de datos de observaciones, sin entrar en la faz especulativa de explicar los fenómenos, hecho que está reservado a una segunda parte, en preparación, con un mayor número de observaciones y análisis a efectuarse entre los años 1969 y 1970.

INTRODUCTION

Since 1965, when Scientific Station Almirante Brown was established by the Instituto Antártico Argentino, an extensive program on oceanographic observations has been developed in the area of Paradise Harbor, a dependency of the Southern Ocean, adjacent to the station. Paradise Harbor is situated in lat. 64°53'S and long. 62°52'W, east of Gerlache Strait, from which is separated by the islands Bryde and Lemaire that give base three approaches.

All these characteristics contribute to a bay of quiet waters, easy to navigate with a small boat, and with more consistent oceanographic processes than in the extensive water masses located westward. Observations in 1965 and 1967 were carried out mainly on the surface while in 1968 more attention was devoted to deep waters. The present paper is a compilation and a report on data. It represents the first part of an oceanographic research program dedicated to explain the physical and chemical phenomena that take place in these antarctic waters.

ARRANGEMENT OF OBSERVATIONS

As was mentioned before, the oceanographic observations performed during the three previous years specifically on surface waters and related to temperature and salinity variations. Additionally, transparency determinations were carried out in 1967 as well as analysis of dissolved oxygen and determinations of pH and total alkalinity on surface waters. During 1968 the activity was centered in deep waters around the boundary of 100 meters depth, the maximum allowed by the small boats available at the station. The

Figure 1 Paradise Harbor. Location of the scientific station Almirante Brown

same year, determinations of dissolved oxygen, phosphorus content in phosphates, and nitrogen content in nitrites were added to the previous work. Transparency determinations were also performed and continuity was pursued regarding analysis on surface salinity, pH and total alkalinity. Deep water sampling was carried out at different levels on surface and at depths of 5, 10, 15, 20, 25, 50, 75 and 90 meters. Each complete sampling operation is called a "series", and accordingly nine series are considered as follows:

1st series:	27/XII/67
2nd series:	16/I/68

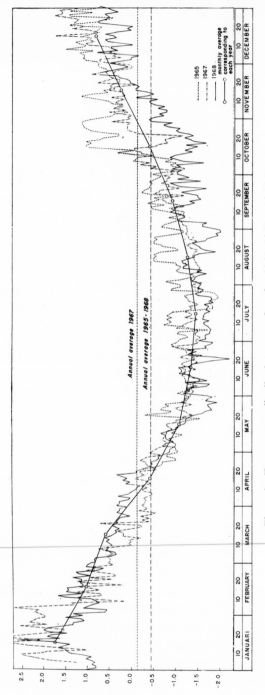

Figure 2 Temperature distribution on surface for 1965, 1967, and 1968. Daily average (based on three-hourly reports), annual average, and average corresponding to each year

3rd series:	7/II/68
4th series:	4/III/68
5th series:	19/IV/68
6th series:	24/V/68
7th series:	24/VI/68
8th series:	25/VII/68
9th series:	8/IX/68

It should be noted that the series collected on March 4th, 1968 is a comparison between samples from Paradise Harbor and Gerlache Strait.

TEMPERATURE DISTRIBUTION ON SURFACE

The temperature records from surface waters correspond to the years 1965, 1967 and 1968. Temperature rates were taken using thermometers specially designed for sea water (SIAP type) by carrying out three measurements daily at 0.00; 12.00 and 18.00 Z hours along the coast and in sea water areas off the station and at depths of 20–30 cm.

Through three years observations a normal evolution is noted showing a maximum in January and a minimum in July. The temperature recordings in 1965 start in April (20th) and they were carried out throughout the years 1967 and 1968. The absolute maximum was 2.8°C in 1967 and 2.63°C in 1968. Both values were taken by mid-January. On the other hand, the absolute minimum was -1.96°C ending July 1965; -1.58°C in early August and -2.2°C in late June 1968. The temperature evolution during this period is standard regarding waters which undergo warming in summer and cooling in winter, presenting no abrupt variations. It is interesting to notice that during that time there was no evidence of water freezing in Paradise Harbor. Some marked daily variations could be due to the tidal regime in the area, as well as pack ice accumulation.

ANNUAL DISTRIBUTION OF TEMPERATURE

Temperature distribution in the water masses related to the year 1968 is plotted in Figure 3, showing isotherms down to a depth of 90 meters between December 1967 and December 1968.

A years data were completed by means curves for which it was assumed that thermal conditions at the beginning of summer 1967 could be repeated almost equally in the summer of 1968. The most remarkable variations shown

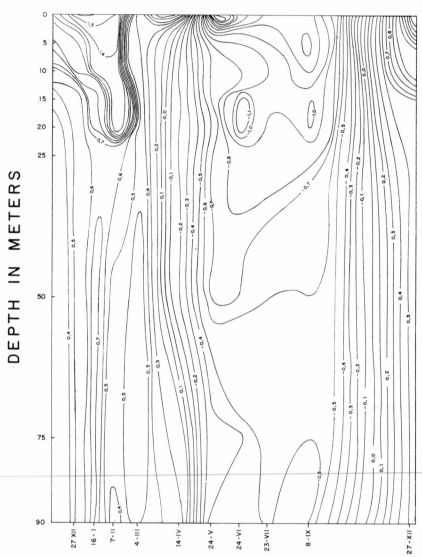

Figure 3 Annual distribution of temperature (at depths from 0 to 90 meters)

by the analysis of the isotherms are indicated by a quick flow of cold water entering the upper layer (depth about 25 meters) that occurs during the second half of February. On the other hand, the layers beneath this level undergo a slower cooling process. In winter, some quick cooling processes take place in the surface layer, between depths of 0 and 5 meters. From late May to September the temperature remains almost constant at each level, except for the layer between 15 and 20 meters which undergoes more cooling, ending in June. Spring warming should take place homogeneously and easily from the surface to the bottom.

SEA WATER SALINITY

Salinity Evolution on surface

The values corresponding to salinity of surface waters for the years 1965, 1967 and 1968 are plotted in Figure 4. The records for the year 1965 cover the period between April and October while in 1967 from May to December, and in 1968 from January to September. The highest salinity was $33.96^o/_{oo}$ recorded in October 1965, $34.87^o/_{oo}$ in late July 1967, and $34.3\ ^o/_{oo}$ in early August 1968. Although the curves are not standard, and do not correspond in maximum and minimum, it can be assumed that there is a general tendency to an increased salinity during the autumn-winter season, and a decrease in the spring. These variations take place to an extent which does not exceed $1^o/_{oo}$ of the mean salinity value, neglecting the few peaks which do not show this condition and do not affect the general evolution. The annual averages, calculated for surface samples are as follows:

Year 1965: $33.62^o/_{oo}$
Year 1967: $33.72^o/_{oo}$
Year 1968: $33.82^o/_{oo}$

Annual distribution of Salinity

The curves cover the whole year 1968, and are related to the layers at depths ranging from 0 to 90 meters. The values were extrapolated according to the same assumptions made in the previous case. These curves are plotted in Figure 5.

It can be noted that in general the most remarkable changes occur within the first 25 meters of depth, although it is interesting to note an abrupt

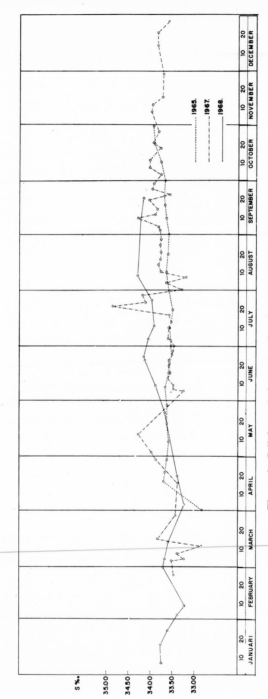

Figure 4   Salinity variation on surface for 1965, 1967, and 1968

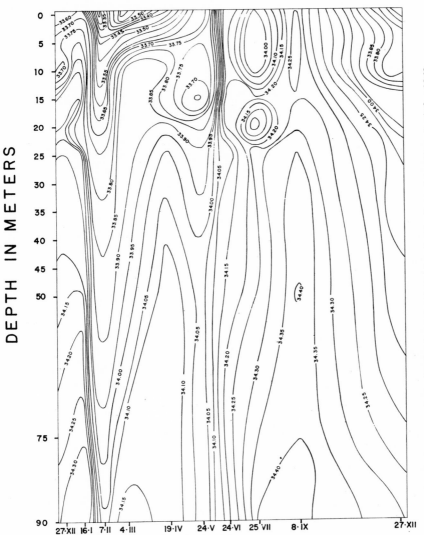

Figure 5   Annual distribution of salinity (at depths from 0 to 90 meters) for 1968

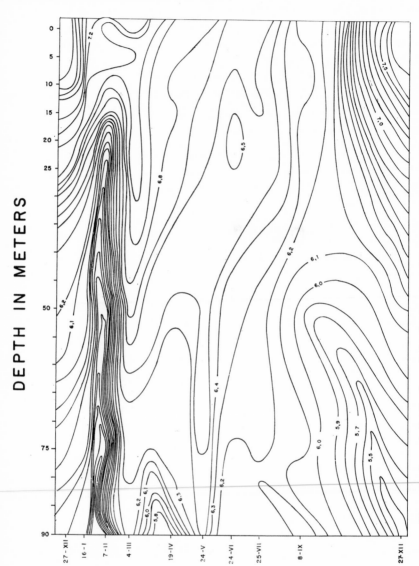

Figure 6　Annual distribution of dissolved oxygen (at depths from 0 to 90 meters) for 1968

erruption of low salinity water in February. This has an influence down to a depth of 90 meters. In March, the layer of surface waters (around 25 meters) shows an increase in salinity, which continues slowly in April. In June, another flow, highly saline, errupts into the same layer for a short time, and is followed by a decrease which is noticeable between depths of 5 and 20 meters. In August there is a new increase in salinity and then a decrease which is evidently related to Spring. After March, the evolution is quite normal in deep waters, ranging from 25 to 90 meters, that is, a progressive increase in salinity in autumn-winter and a gradual decrease in spring.

ANNUAL DISTRIBUTION OF DISSOLVED OXYGEN

The annual distribution of oxygen concentrations for the year 1968 are plotted in Figure 6. True values given till September were used and then, the curves were extrapolated from September onward, making the same assumptions as for salinity and temperature.

From the analysis of the diagrams it is concluded that the major summer variations are restricted to levels of depth between 15 and 90 meters, and the rest of the year, though lower, are confined to deep levels (50 to 90 meters). Generally, in winter, the tendency in the upper levels is to decrease concentration in a small average and very slowly until spring; then, the increase develops more rapidly, particularly in the upper layer (0 to 15 meters). Finally, it should be considered that the concentration of dissolved oxygen in this area and extensively in antarctic waters, are higher in a broad sense than those reported for other oceans. This characteristic is evident throughout the year.

SEA WATER TRANSPARENCY

The data available correspond to the years 1967 and 1968 and are plotted in Figure 7.

The transparency was taken as proportional to the amount of organic matter suspended in sea water. Through the diagram it can be noted that in the warmest months, January and February, the transparency reaches a depth of 6 meters with slight fluctuations, and in early March it increases and is evident down to depths of 17–18 meters. Maximum transparency in

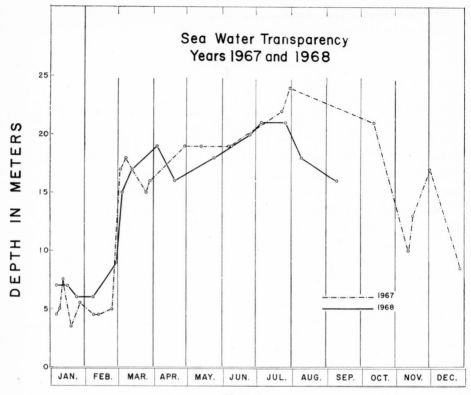

Figure 7

both years is achieved at the end of July and in early August, corresponding to 24 meters in 1967 and to 21 meters in 1968. Since then, it tends to decrease gradually. Records in 1968 end in September, but those for 1967 register an abrupt decrease from 21 to 20 meters in the period between late October and the beginning of November, showing an increase (17 meters of depth) in late November and then a true decrease during December, reaching a depth of 8.50 meters. As an indication of biological activity, the transparency would suggest that:

1. There are two well defined water masses, one corresponding to the spring-summer season and the other to autumn-winter.

2. Changes in both water masses occur abruptly, from the last days of February to early March in the former, while in the latter they take place in early November.

These changes in the amount of organic matter are related to salinity, temperature, pH, total alkalinity, and dissolved oxygen and nutrients. These last two elements fluctuate due to biological activity.

pH VARIATION

The values for pH variation on the surface during 1967 and those taken at different levels during 1968 are recorded in Table 1. While the period considered in 1967 was short, it is worthwhile to note that the minimum value occurs in May; then a marked increase is observed, ceasing early in June, following a slight decrease and then an increase toward the first days of July. From then on, apparently, a true decrease begins. The analysis in 1968 revealed that, excepting the surface layer and down to 5 meters, the variation recorded in values given for depths down to 90 meters is slight and it ranges from pH 7.80 to 7.90, showing a minor decrease toward mid-April, a slight increase ending May and then, remaining almost constant from May to September. Surface layers down to 5 meters are included in these variations. In surface layers and down to 5 meters depth, a marked increase occurs in March for surface values and in April for those related to 5 meters. The minimum pH recorded for these levels is 7.70.

TOTAL ALKALINITY VARIATION

From the surface rates recorded during 1967, it can be concluded that these are stable from April to mid-May, and then they turn to decrease until June. In mid-winter, a sinuous increase takes place until July, date of the last record. The tendency suggests that the process is likely to increase. The total alkalinity variation related to the nine series in 1968, is recorded in Meq/l in Table 1.

Throughout the period of observations, in general the most stable values are those given for layers at depths of 15, 20 and 25 meters. On the other hand, they range irregularly from 0 to 10 meters, as well as from 50 to 90 meters. Values are given, related to the layer from 0 to 5 meters and similarly for those between 50 and 90 meters. Since July, stability is observed in values related to all layers. Maximum and minimum values are given in the same series (24/V/68) corresponding to 2.29 Meq/l for 25 meters and 0.85 Meq/l for 50 meters. The lowest difference corresponding to this series is achieved in July, when from surface layers and down to a depth of 90 meters the difference is 0.1 Meq/l.

Table 1 Variation of pH and total alkalinity
Years 1967 and 1968

Sample	Date	Depth (m)	pH	Total Alkalinity (Meq/l)
15	28/IV/67	0	8.23	1.758
16	28/IV/67	0	—	2.025
17	28/IV/67	0	—	2.226
18	28/IV/67	5	—	1.937
19	2/V/67	7	8.40	1.975
20	2/V/67	0	—	1.999
21	9/V/67	0	8.33	2.266
22	9/V/67	0	—	2.101
23	9/V/67	0	—	1.705
24	12/V/67	0	8.13	2.044
25	12/V/67	0	—	—
26	5/VI/67	0	8.26	1.567
27	6/VI/67	0	8.72	1.755
28	6/VI/67	0	8.72	1.567
29	6/VI/67	0	8.52	1.182
30	8/VI/67	0	8.58	1.669
31	12/VI/67	0	8.21	1.589
32	15/VI/67	0	8.77	1.522
33	19/VI/67	0	8.42	1.705
34	22/VI/67	0	8.31	1.975
35	27/VI/67	0	8.52	1.669
36	30/VI/67	0	8.47	1.795
37	3/VII/67	0	8.34	1.911
38	3/VII/67	0	8.63	1.710
39	6/VII/67	0	8.48	1.839
40	10/VII/67	0	8.35	2.104
41	13/VII/67	0	8.40	1.975
42	17/VII/67	0	8.21	2.031
41	28/II/68	0	8.08	1.88
50	4/III/68	0	7.82	2.09
46	4/III/68	5	7.82	2.16
45	4/III/68	10	7.82	2.07
44	4/III/68	15	7.82	2.07
43	4/III/68	20	7.81	2.07
42	4/III/68	25	7.81	2.16
47	4/III/68	50	7.76	2.13
48	4/III/68	75	7.76	2.16
49	4/III/68	90	7.78	2.17
59	4/III/68	0	7.88	2.04

Table 1 (*cont.*)

Sample	Date	Depth (m)	pH	Total Alkalinity (Meq/l)
51	4/III/68	5	7.88	2.09
52	4/III/68	10	7.86	2.11
53	4/III/68	15	7.86	2.09
54	4/III/68	20	7.88	2.09
55	4/III/68	25	7.88	2.16
56	4/III/68	50	7.86	2.12
57	4/III/68	75	7.87	2.09
58	4/III/68	90	7.87	2.17
60	13/III/68	0	7.60	1.67
61	3/IV/68	0	7.73	1.91
70	19/VI/68	0	7.78	1.76
62	19/IV/68	5	7.58	1.19
63	19/IV/68	10	7.85	2.09
64	19/IV/68	15	7.85	1.98
65	19/IV/68	20	7.79	1.98
66	19/IV/68	25	7.85	1.98
67	19/IV/68	50	7.79	2.13
68	19/IV/68	75	7.79	1.19
69	19/IV/68	90	7.79	2.09
71	24/V/68	0	7.85	1.33
72	24/V/68	5	7.84	1.76
73	24/V/68	10	7.84	1.19
74	24/V/68	15	7.90	1.98
75	24/V/68	20	7.89	1.98
76	24/V/68	25	7.89	2.09
77	24/V/68	50	7.89	0.85
78	24/V/68	75	8.89	2.29
79	24/V/68	90	7.89	1.98
80	8/VI/68	0	7.86	2.14
84	24/VI/68	0	7.91	1.98
87	24/VI/68	15	7.91	1.98
88	24/VI/68	20	7.91	1.98
89	24/VI/68	25	7.90	1.98
90	24/VI/68	50	7.90	2.09
91	24/VI/68	75	7.90	2.09
92	24/VI/68	90	7.90	1.19
93	25/VII/68	0	7.91	2.09
94	25/VII/68	5	7.85	2.09
95	25/VII/68	10	7.90	2.09
96	25/VII/68	15	7.90	2.09

Table 1 *(cont.)*

Sample	Date	Depth (m)	pH	Total Alkalinity (Meq/l)
97	25/VII/68	20	7.90	1.99
98	25/VII/68	25	7.90	1.99
99	25/VII/68	50	7.90	1.99
100	25/VII/68	75	7.90	1.99
101	25/VII/68	90	7.85	1.99
103	8/IX/68	0	7.91	2.09
104	8/IX/68	5	7.91	2.09
105	8/IX/68	10	7.85	2.09
106	8/IX/68	15	7.79	2.09
107	8/IX/68	20	7.85	1.46
108	8/IX/68	25	7.85	2.09
109	8/IX/68	50	7.85	1.99
110	8/IX/68	75	7.90	1.99
111	8/IX/68	90	7.85	1.99

PHOSPHATES AND NITRITES

In 1968, phosphates were determined in five series covering the months from April to September. They are recorded in Table 2 and are expressed in Ma P—PO_4/m³.

From April till early June a strong increase is observed at all levels, excepting the layer at a depth of 50 meters. The major increases occur at levels of 10, 20 and 25 meters. In June the concentration at all levels decreases, excepting on the surface where it increases until early July. Regard-

Table 2 Variation of phosphates and nitrites
Years 1968

Sample	Date	Depth (m)	Phosphates (mg/m³)	Nitrites (mg/m³)
61	3/IV/68	0	68.82	—
70	19/IV/68	0	68.82	—
62	19/IV/68	5	78.12	—
63	19/IV/68	10	59.83	—
64	19/IV/68	15	72.54	—

Table 2 *(cont.)*

Sample	Date	Depth (m)	Phosphates (mg/m^3)	Nitrites (mg/m^3)
65	19/IV/68	20	78.12	—
66	19/IV/68	25	81.53	—
67	19/IV/68	50	88.66	—
68	19/IV/68	75	59.83	—
69	19/IV/68	90	68.82	—
71	24/V/68	0	88.66	7.70
72	24/V/68	5	96.10	7.70
73	24/V/68	10	111.91	8.54
74	24/V/68	15	88.66	7.70
75	24/V/68	20	123.38	7.70
76	24/V/68	25	116.56	7.00
77	24/V/68	50	76.88	7.00
78	24/V/68	75	88.66	7.70
79	24/V/68	90	88.66	8.54
84	24/VI/68	0	98.11	15.05
87	24/VI/68	15	84.94	8.85
88	24/VI/68	20	79.67	8.85
89	24/VI/68	25	79.67	8.85
90	24/VI/68	50	74.24	7.98
91	24/VI/68	75	84.94	7.98
92	24/VI/68	90	74.24	7.98
93	25/VII/68	0	70.99	5.08
94	25/VII/68	5	51.77	6.00
95	25/VII/68	10	62.31	4.16
96	25/VII/68	15	62.31	3.23
97	25/VII/68	20	62.31	5.08
98	25/VII/68	25	62.31	3.23
99	25/VII/68	50	51.77	4.16
100	25/VII/68	75	51.77	4.16
101	25/VII/68	90	51.77	3.23
103	8/II/68	0	38.19	—
104	8/II/68	5	38.19	—
105	8/IX/68	10	38.19	—
106	8/IX/68	15	38.19	—
107	8/IX/68	20	33.70	—
108	8/IX/68	25	33.70	—
109	8/IX/68	50	44.92	—
110	8/IX/68	75	44.92	—
111	8/IX/68	90	49.41	—

ing the layer at 30 meters, it decreases constantly from April to September. In September, minimum values are recorded in all layers ranging from 1.09 to 1.59 at/μg P—PO$_4$/l. The highest rate was obtained ending May at levels between 10 and 25 meters, ranging from 3.61 to 3.98 at/μg P—PO$_4$/l. Nitrite determinations are restricted to three series which cover May, June and July 1968. The corresponding values are shown in Table 2 and are expressed in Mg N—NO$_2$/m^3. A slight increase in concentration related to May and June is observed in the three series, followed by a marked decrease ending July.

The major variation is recorded on surface reaching a maximum of 1.055 at/μg-NO$_2$/l and a minimum of 0.353 at/μg-NO$_2$/l in July. The variation related to the other layers does not exceed 0.10 at/μg N—NO$_2$/l. It is interesting to note that the layer at 90 meters presents a concentration that decreases from May to June. These particularly high records are in accordance with those reported for the area by S. A. El-Sayed in 1968.

CONCLUSIONS

From the preceding it can be inferred that:

1. A remarkable change in the physico-chemical behavior of sea water in Paradise Harbor was observed in the period between late February and mid-March.

2. The change takes place abruptly, and delimits a summer mass of water from a winter one.

3. Variation involves all chemical aspects studied and included a marked change in temperature.

4. Since then, the evolutive processes that occurred suggest a tendency to uniformity in the chemical composition of the water masses flowing from the layer which would be the apparent source of the components considered.

In the autumn-winter season, sea water increases in salinity values while its oxygen concentration decreases slowly and the phosphate and nitrite contents augment, particularly in surface layers. Deep waters (at depths around 50 meters) which are likely to support a poor biological activity, do not fulfil strictly such normal condition, and only in mid-winter (July–August), when the remaining activity disappears (or is very low) the components involved would enter into a process of diffusion and flow toward the whole mass of water, being this the reason of the decrease that is evident in general.

References

BALECH, E., EL SAYED, S. Z., HASLE, G., NEUSHUL, M. and ZENEVELD, J. S. (1968). *Primary Productivity and Benthic Marine Algae of the Antarctic and Subantarctic.* Folio Series. Folio 10. American Geographical Society. New York.

MIECZYSLAW, O. (1962). *The Determination of Chlorinity by the Knudsen Method.* Woods Hole Oceanographic Institution. New York.

SERVICIO DE HIDROGRAFIA NAVAL. (1958). *Manual de Instrucciones para Observaciones Oceanográficas. H 601.* Ed. S. H. N. Buenos Aires.

SERVICIO DE HIDROGRAFIA NAVAL. (1959). *Química del Agua de Mar. H 604.* Ed. S.H.N. Buenos Aires.

STRICKLAND, J. D. H. and PARSONS, T. R. (1965). *A Manual of Sea Water Analysis.* Bulletin No 125. Fisheries Research Board of Canada. Ottawa.

On the mixing of water masses

LUIS R. A. CAPURRO

Texas A and M University
Department of Oceanography
College Station, Texas, U.S.A.

Abstract

Main physical processes which tend to eliminate concentration differences in the properties of geophysical fluids are: a) those related to the movement of large water masses, and b) those produced by the diffusion of those concentrations, which are basically of a turbulent nature.

The effectiveness of mixing processes in the ocean is clearly demonstrated by the stationary vertical distribution of temperature, salinity, oxygen and other properties. The main approach of this work centers on the analysis of these stationary distributions.

In physical oceanography the mixing of water masses is studied with the aid of the T-S diagram. This is commented upon in the present state of the knowledge of general oceanic circulation, responsible for the large scale mixing, as well as concepts on turbulent mixing in the ocean. On the basis of a study carried out by Munk (1966) on the vertical distribution of variables in the Pacific Ocean, different possible modes of horizontal and vertical mixing are analyzed, such as: a) boundary mixing, b) thermo-dynamical mixing, c) shear mixing, and biological mixing.

Considerations on the research work carried out by Munk (1966) in the South Pacific, and that of Lynn and Reid (1968) in the analysis of data from the abyssal Atlantic are made.

Resumen

Los principales procesos físicos que tienden a eliminar diferencias en concentraciones de propiedades en los fluidos geofísicos son: a) los relacionados con los desplazamientos de grandes volúmenes de agua que llevan consigo todas las concentraciones arriba mencionadas y que incluyen

movimientos horizontales y verticales, y b) los producidos por la difusión
de esas concentraciones, las cuales son fundamentalmente de carácter
turbulento y que ocurren en una gama muy amplia de escalas en el sentido
horizontal y vertical. La efectividad de los procesos de mezcla en el océano
está claramente demostrado por la constancia de las distribuciones verticales de la temperatura, salinidad, oxígeno y otras propiedades.

En oceanografía física, la mezcla de masas de agua se estudia con la
ayuda del diagrama T-S. Se comenta en el estado actual de los conoci-
mientos de la circulación general oceánica, responsable del mezclado en
gran escala y conceptos sobre mezcla turbulenta en el océano. Sobre la
base de un estudio realizado por Munk (1966) acerca de la distribución
vertical de variables en el Océano Pacífico se analizan distintos posibles
modos de mezcla horizontal y vertical tales como a) mezcla en las inter-
faces, b) mezclado termodinámico, c) mezclado por gradiente vertical de
la velocidad horizontal y mezclado biológico.

Se mencionan ligeramente los trabajos realizados por Munk (1966) en
el Pacífico Sur y los de Lynn y Reid (1968) en el análisis de los datos del
Atlántico abisal.

One of the complexities of the geophysical fluids is that the continual
exchange of energy and matter across the boundaries of the ocean prevents
the establishment of a complete thermodynamical equilibrium and is
responsible for the also continuous change in the concentration of momentum, heat, and salt.

The main natural physical processes that try to destroy the difference
in concentrations are of two kinds, namely: the advective mode, which
involves large-scale displacement of waters carrying with them the above
concentrations, and the diffusive mode, where equalization of concentrations
takes place without any overall transport of water. The advective processes
are related to the tri-dimensional field of motion in which ocean currents at
all depths and the slow vertical movements of water masses (upwelling and
downwelling) are included. Diffusive exchange is produced by turbulent
mixing on a very wide range of scales in both vertical and horizontal directions. Molecular processes are of very little importance in large-scale natural
systems.

These two modes of mixing are in continuous operation and exist on a
large variety of scales and intensities, which may lead one to think that a
thorough mixing would result so that the water of the oceans would be
homogeneous. We know that this is not the case because other processes
bring about differences between water masses which are maintained by the
layering or stratification of the sea. The thermal structure of the water

column, the vertical gradient of salt content, and the relative "constancy of composition" provide eloquent information about the effectiveness of horizontal mixing activity and to the resistance of the World Ocean to vertical exchanges.

Before getting more involved in the analysis of the mixing processes in the World Ocean, it is convenient to define the problem to be discussed. This problem is as follows:

1. The differential heating of the oceans, with a surplus of energy stored in low latitudes and a deficit of energy in high latitudes, creates such a temperature gradient that the poleward flow of heat from the equatorial regions required to take care of the differences in temperature is beyond the capabilities of conductive processes. So, fluid motions develop and important amounts of heat and other concentrations are transported by them. The system of fluid motions is known as the general circulation of the World Ocean and it is responsible for the large-scale mixing activity. This circulation is of a turbulent nature.

2. The interaction processes through the surface and lateral boundaries of the ocean generate changes in concentrations of momentum, matter and energy. They also create conditions for the diffusive exchange of the above concentrations. This is known as "boundary mixing". The strong mixing activities at the boundaries are transmitted to the interior of the fluid along surfaces of potential density.

3. In spite of the continuous change of concentrations and the mixing activity that tries to destroy them, there is a definite vertical concentration of density, heat, salt, and gas which are maintained by different mixing processes which are responsible for this stationary state of the distributions.

The problem to be discussed is to describe the theories which explain the facts mentioned in (3).

In physical oceanography the mixing of water masses is studied with the aid of the temperature-salinity diagram. This concept, which implies that the temperature is a universal function of salinity, applies to many large regions of the ocean. A logical approach to the mixing problem in a geophysical fluid, like the water of the oceans, is to explain the distribution and permanency of the temperature and salinity distributions. When data are available, it is convenient to extend the study to other concentrations, such as oxygen and radioactive elements.

It is convenient to clarify certain basic concepts about certain characteristics of sea water as we find it in the ocean. These are as follows:

1. The waters of certain regions of the World Ocean have a characteristic relationship of salinity and temperature in both a horizontal and vertical extent. The stratification in the bodies of water is stable, meaning that the density increases with depth. That characteristic relationship is known as T-S relationship, and when it is plotted, it becomes the T-S diagram.

2. Because of this relationship, it is possible to distinguish bodies of water as "water masses". The T-S diagram reflects the processes that conditioned the water mass when it was at the surface in contact with the atmosphere and the mixing processes that occurred within the interior of the ocean.

3. The density of sea water can be changed at the surface of the oceans by heating or cooling, by evaporation or precipitation, or by the freezing or melting of ice. Mixing of two water types also affects density and even the mixture of water types of the same density, but different temperature and salinity will produce water of higher density (caballing). The mixture of water types is known as thermodynamic mixing.

It is time now to discuss in some detail the first mode of mixing activity, the one related to the fluid motions.

LARGE SCALE MIXING

The large scale mixing generated by advective processes depends entirely on the oceanic circulation, the general features of which can be explained as a result of two combined modes of circulation: the wind-driven circulation and thermohaline circulation.

The first type of distribution of currents resembles closely the pattern of the planetary surface wind field. Taken as a whole, the wind-spun vortex is asymmetrical and circulates around a point displaced considerably to the west of the center of each of the major oceans in the tropical and temperate latitudes of the earth. Figure 1 depicts these features in the Atlantic Ocean. These vortices are characterized by the fact that the poleward circulation along the western side of each ocean tends to become concentrated in a narrow and fast flow as it approaches a solid continental barrier. These are known as "western boundary currents", and the well-known examples are the Gulf Stream of the North Atlantic, the Brazil Current of the South Atlantic, and the Kuroshio or Black Current in the North Pacific. These

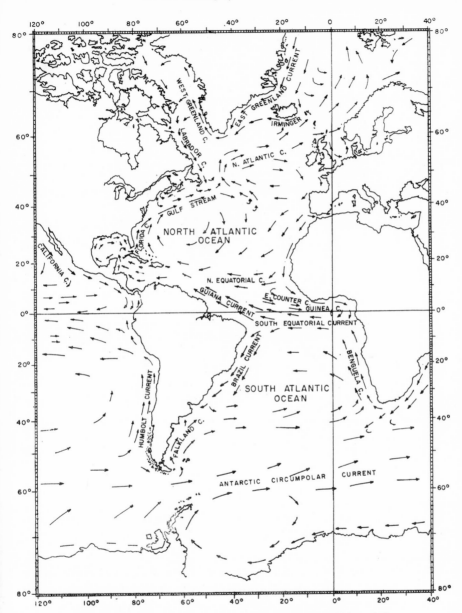

Figure 1   Main ocean currents of the Atlantic Ocean. (Adapted from *Encyclopedia Britannica World Atlas*, Chicago)

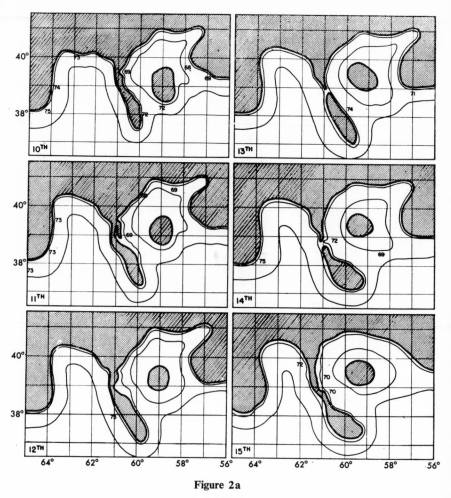

Figure 2a

Figure 2a and b   The evolution of eddy "Edgar" during 10–22 June 1950.
(From F. C. Fuglister and L. V. Worthington, 1951, *Tellus*, **3**, 1

currents display a strong tendency to meander and detach sizeable eddies which act as a frictional brake by large-scale eddy diffusion. The wavelength of the meanders ranges from about 80 to 250 nautical miles. The best studied life history of an eddy has been done by Fuglister (1951) during the multiple ship operation CABOT in the Gulf Stream. Figure 2 shows a sketch of the development of such a feature. The tendency to meander has

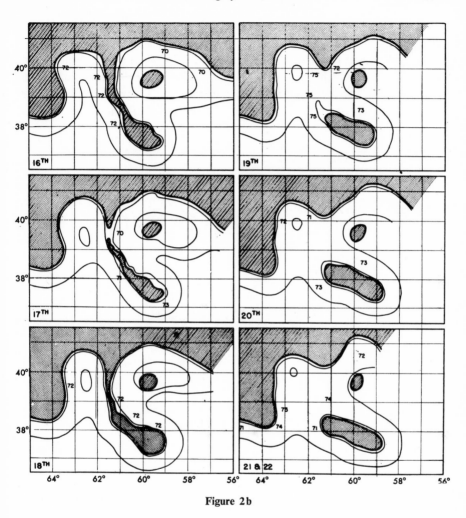

Figure 2b

been detected in the Brazil Current and has been frequently observed in the meteorological satellite information.

The thermohaline circulation or convective circulation maintained by the conditioning of the water at the surface by heating, evaporation and precipitation processes is a slow overturning of the ocean waters produced by the sinking of high-density water in certain oceanic regions (concentrated sources) and slow rising motions everywhere else in the World Ocean (distributed sinks)—Stommel (1957). In a remarkable study, Stommel identifies two principal sources of abyssal circulation. One is located in the Labrador and Irminger Seas, which generates a narrow deep current flowing south along the east coast of North and South America and the other in the shelf of the Weddell Sea in Antarctica, which is the source of another deep western

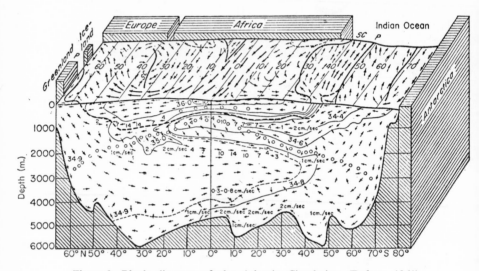

Figure 3 Block diagram of the Atlantic Circulation (Defant, 1961).
Physical Oceanography, Vol. 1. Pergamon Press

boundary current that flows north in the South Atlantic. The circulation around Antarctica in the Southern Ocean plays an important role because of the impressive water volume transport and because of the fact that it is the principal avenue of exchange between the major ocean basins. Figure 3 shows the block diagram of the Atlantic Circulation, and Figures 4 and 5 show the horizontal current fields in the Atlantic Ocean at 800 m and 2000 m according to Defant (1961).

So, the broad features of the general circulation of the World Ocean can be explained as the result of the two modes of motion already explained. A great variety of motions of different sizes and frequencies are included within the general aspects already discussed. This complex general circulation is responsible for the large-scale mixing processes in the World Ocean.

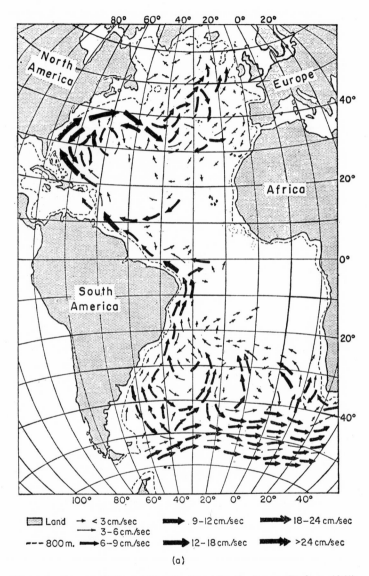

Figure 4 Current field at 800 m in the Atlantic Ocean (Defant, 1961). *Physical Oceanography*, Vol. 1. Pergamon Press

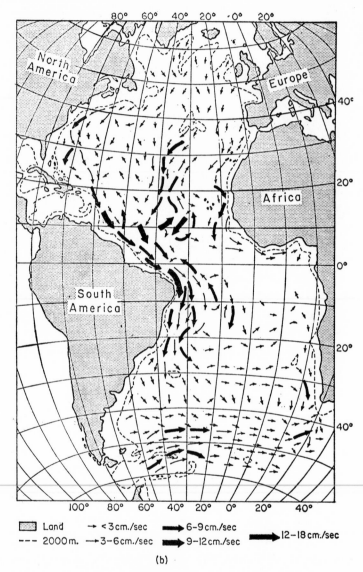

Figure 5 Current field at 2000 m in the Atlantic Ocean (Defant, 1961). *Physical Oceanography*, Vol. 1. Pergamon Press

DIFFUSIVE MIXING

Diffusive effects are produced by turbulent mixing on a very wide range of scales, both in the vertical and horizontal directions. Turbulent mixing is much greater than molecular diffusion, and it is a common process in the natural oceanic environment. It is generated because of the turbulent nature of the flow of water in the sea. The turbulent motion is usually referred to as "eddies", without any implication of structure or size. They cover a wide spectrum in both aspects. The efficiency of turbulent mixing in the vertical direction is much smaller than the mixing in the horizontal direction. This is explained as the result of two factors—the large ratio of width to depth of the ocean and the influence of the normal vertical stratification of the ocean, which reduces considerably vertical turbulence without affecting horizontal turbulence.

Turbulent vertical mixing is mainly generated by two different processes: (a) wind stirring at the ocean surface through wave motion and surface drift (boundary mixing) and (b) vertical shear in the horizontal currents, which may be produced by different factors, such as bottom friction, internal tidal waves, and internal planetary waves.

In shallow water, in addition to wind-generated turbulence at the surface, bottom friction creates turbulence near the bottom in such a way as to maintain an active mixing along the whole column of water. This occurs off the coast in Southern Argentina where the tidal currents are greater than 150 cm/sec. In these shallow regions of intense turbulence, the pycnocline formation is prevented by the strong mixing.

In his analysis of the vertical distributions in the interior Pacific Ocean, Munk (1966) obtained a value for the vertical eddy diffusivity K 1 cm^2 sec^{-1}. In his efforts to reconcile the processes responsible for vertical mixing in the deep interior with physical considerations, he considered four interesting modes of vertical mixing—boundary mixing, thermodynamic mixing, shear mixing, and biological mixing.

Boundary mixing

In his work on the North Atlantic waters, Iselin (1939) reasoned that the water attains its T-S characteristics by surface processes, and these characteristics are transmitted into the interior along surfaces of constant potential density (isentropic mixing). The remarkable resemblance of the vertical T-S relation with the surface horizontal T-S relation was found in the North

Atlantic and is also present in the Pacific. This surface mixing extends down to a depth of 1000 m.

If the surface boundary is instrumental in establishing the properties of the first km, it is possible that the lateral boundaries may also be instrumental in determining the distribution to the deeper waters. One can think of strong mixing along the ocean boundaries and islands due to current shear and internal wave breaking coupled with an effective communication into the interior along potential density surfaces.

Thermodynamic mixing

This type of mixing is related to the instabilities which can develop in a liquid which is hydrostatically stable as a result of the different rates of diffusion of its potential temperature and salinity (Stommel *et al.*, 1956; Stern, 1960; Turner and Stommel, 1964; Turner, 1965; Turner, 1967; and Stern, 1968).

In Stern's experiments, warm, salty water was floated over heavy, cold, fresh water. Vertical convective filaments were present across the interface within an hour (salt fingers). In Turner's experiments, relative light, cold fresh water floated over heavy, warm, salty water. Convective turbulence and progressive layering resulted. Although some of the experiments' conditions are not repeated in the ocean, the possibility of convection in the deeper layers on each side of an interface containing salt fingers has been suggested to take place in the Mediterranean where layering under the Mediterranean outflow in the Atlantic Ocean has been observed.

Foster (1968) and personal communication (1969) has done interesting laboratory experiments on the haline convection induced by the freezing of sea water in an effort to understand the formation of the Antarctic Bottom Water. He has frozen sea water at different rates and found that haline convection took the form of long vertical filaments with a horizontal spacing in good agreement with the predictions of the linear theory.

Another aspect of thermodynamic mixing is the "caballing" process mentioned earlier.

Shear mixing

This type of instability is generated into a statically stable fluid in the presence of vertical shear when certain conditions are fulfilled. Munk analyzed the conditions when instability occurs on the basis of the Väisälä frequency which measures the static stability and the shear produced by internal waves of tidal frequency and internal planetary waves. In both cases, instability is achieved if internal wave energy is concentrated among very high modes with typical velocities of the order of 10 cm-sec^{-1} and corresponding wavelengths at tidal frequencies of 5–10 km and 4 km for the planetary waves. This requirement is too severe for both frequencies. However, one possible source of high mode internal wave energy is the conversion of long surface waves to internal waves over an irregular sea bottom.

Biological mixing

This mixing has been studied in relation to the horizontal and vertical distributions of dissolved oxygen. A consumption figure was obtained which is broken down in different mixing activities, such as flow perturbation by swimming, migration of zooplankton and nekton and differential feeding.

Lateral mixing

The horizontal mixing occurs in a very broad spectrum of horizontal motions. The eddies detached from the core of the wind-driven flow, particularly from the western boundary currents, are of different sizes, and the effective mixing takes place by the cascade type of breaking off in eddies of smaller sizes. The horizontal turbulent exchange coefficient K_e depends, in these cases, on the scale of the phenomenon and the growth of K_e follows different laws as the scale of the phenomenon varies. Ozmidov (1968) devised a model applicable to average oceanic conditions on the basis of the distribution of kinetic-energy density for motion in the ocean on various scales.

Stern (1967) has shown that in a water mass with large-scale variations of temperature and salinity on isopycnal surfaces and positive vertical salt gradient, instabilities develop which transport salt along the isopycnal surfaces. This implies that any large-scale variation of the temperature and salinity gradients on isopycnal surfaces tends to be removed by instabilities in the same way that super-adiabatic gradients in the free region of a heated fluid tend to be removed by vertical convection.

A very interesting study on the vertical distributions in the interior Pacific (excluding the top and bottom kilometer) was made by Munk (1966). On the basis of steady conditions, he found that the vertical distributions of T, S, ^{14}C, and O_2 can be explained as a simple model involving a constant upward velocity of around 1.2 cm-day^{-1} and eddy diffusivity of around

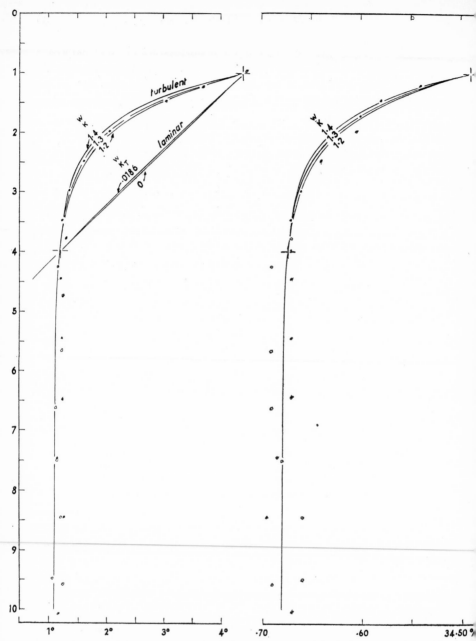

Figure 6 Potential temperature and salinity as functions of depth (km) at station *Snellius* 1930: No. 262,9° 41′ N, 126° 51′ E, (closed circles) and *Galathea* 1951: No. 433,9° 51′ N, 126° 51′ E, (open circles). Curves labeled *w/k* (in units km^{-1}) are based on equations (1) and (2) for turbulent and laminar diffusion, respectively. (From Munk, 1966, *Deep Sea Research*, Vol. 13. Pergamon Press)

1.3 cm²-sec⁻¹. These vertical distributions are consistent with the corresponding south-north gradients in the deep Pacifc, provided there is an average northward drift of at least a few millimeters per second. He interpreted the inferred rates of upwelling on the basis of the formation of bottom water

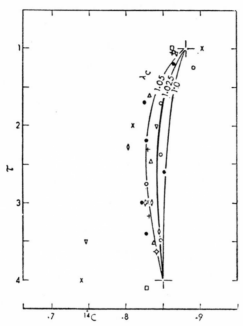

Figure 7 ¹⁴C (relative units) as function of depth (km) (from Bien, Rakestraw and Suess, 1965). Symbols represent various Pacific stations. Curves are based on equation (6) for $y = 3.3$, and for indicated values of λ_c. (From Munk, 1966, *Deep Sea Research*, Vol. **13**. Pergamon Press)

in Antarctica with relative success but had difficulties in the interpretation of the inferred diffusion in terms of known processes. No definite improvement can be made on his model until the processes related to diffusion and advection are better understood.

Figures 6, 7 and 8 show the vertical distributions used by Munk to obtain the values reported for his model.

Lynn and Reid (1968) examined the abyssal properties in the World Ocean with newer data and extended the principles of isentropic analysis to greater depths. They found the presence of a potential-density maximum above the bottom in the Western Atlantic, which represents fairly well the lower

North Atlantic Deep Water. This maximum makes it possible to trace water masses even farther through the Antarctic Circumpolar Current.

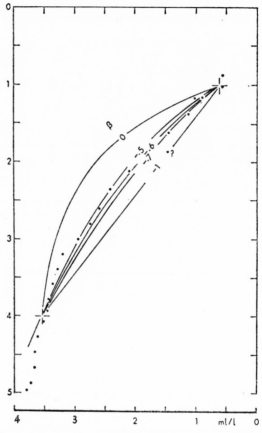

Figure 8 Oxygen in ml/l as function of depth (km) of station *Calcofi* No. 60.190. Curves are based on equation (8) for $\lambda = 3.3$, and for indicated values of β. (From Munk, 1966, *Deep Sea Research*, Vol. 13. Pergamon Press.)

Acknowledgement

Figure 2. Reprinted with permission from F. C. Fuglister and L. V. Worthington, Some Results of a Multiple Ship Survey of the Gulf Stream, Tellus, 1951.

Figures 3, 4, and 5. Reprinted with permission from A. Defant, Physical Oceanography, Volume I, 1968, Pergamon Press.

Figures 6, 7, and 8. Reprinted with permission from Walter H. Munk, Abyssal Recipes, Deep Sea Research, 1966, Pergamon Press.

References

BOWDEN, K. F. (1965). Currents and mixing in the ocean. In *Chemical Oceanography*, pp. 43–72. Edited by J. P. Ryley and G. Skirrow. London: Academic Press.

DEFANT, A. (1961). *Physical Oceanography*. Vol. 1. Oxford, London, New York, Paris: Pergamon Press.

FOSTER, T. D. (1968). Haline convection induced by the freezing of sea water. *Journal of Geophys. Research.* **73**, 1933–1938.

FUGLISTER, F. C., and WORTHINGTON, L. V. (1951). *Tellus*, **3**, 1.

ISELIN, C. O'D. (1939). The influence of vertical and lateral turbulence on the characteristics of the waters at middepths. *Trans. Amer. Geophys. Union.* **7**, 414–417.

LYNN, R. L. and REID, J. L. (1968). Characteristics and circulation of deep and abyssal waters. *Deep Sea Research.* **15**, 577–598.

MUNK, W. (1966). Abyssal recipes. *Deep Sea Research.* **13**, 707–770.

OZMIDOV, R. V. (1968). *Izv. Atmospheric and Ocean Physics*, **4**, 11, pp. 1224–1225. Translated by J. Findlay.

STERN, M. (1960). The "salt fountain" and thermohaline convection. *Tellus.* **12**, 172–175.

STERN, M. E. (1967). Lateral mixing of water masses. *Deep Sea Research.* **14**, 747–753.

STERN, M. E. (1968). T-S gradients on the micro-scale. *Deep Sea Research.* **15**, 245–250.

STOMMEL, H., ARONS, A. B., and BLANCHARD, D. C. (1956). An oceanographical curiosity: the perpetual salt fountain, *Deep Sea Research.* **3**, 152–153.

STOMMEL, H. (1957). A survey of ocean theory. *Deep Sea Research.* **4**, 149–185.

TURNER, J. S., and STOMMEL, H. (1964). A new case of convection in the presence of combined vertical salinity and temperature gradients. *Proc. U. S. Nat. Acad. Sci.* **52**, 49–63.

TURNER, J. S. (1965). The coupled turbulent transports of salt and heat across a sharp density interface. *Inst. J. Heat Transfer.* **8**, 759–767.

TURNER, J. S. (1967). Salt fingers across a density interface. *Deep Sea Research.* **14**, 599–611.

Considerations on the ichthyoplankton in the shelf waters of the Southwestern Atlantic, in front of Argentina, Uruguay and southern part of Brazil

JANINA D. DE CIECHOMSKI

Instituto de Biologia Marina
Mar del Plata, Argentina

SUMMARY

The main results obtained in the present study are:

1) In every cruise anchovy, *Engraulis anchoita* eggs and larvae were found. This indicates that this species reproduces during the whole year. The intensity of its spawning varies according to the season of the year.

2) The frequency of anchovy eggs and larvae in relation to all other species of fish is very high at every season.

3) The high density of the eggs of the anchovy, even when compared with the Peruvian anchoveta, indicates the existence of very large fishery resource of *Engraulis anchoita*.

4) The most important spawning area of the anchovy, located more or less in front of Rio de La Plata, varies in a southern or northern direction according to the season of the year.

5) It was found that the occurrence of anchovy eggs was in positive relation to the occurrence of eggs of other species of fish. In the majority of cases

large numbers of other fish eggs were related to a large number of anchovy eggs.

6) In the majority of cases the large number of anchovy eggs was related to small numbers of zooplankton organisms. It is assumed the big shoals of spawning anchovy can seriously affect the zooplankton populations.

RESUMEN

En el trabajo están presentadas unas consideraciones generales sobre el ictioplancton en las aguas de la plataforma continental del Océano Atlántico frent a la parte norte de la Argentina, hasta el paralelo 47° 55′ S, frente al Uruguay, y frente a la parte sur de Brasil. Los resultados más importantes son los siguientes:

1) Los huevos y larvas de la anchoíta, Engraulis *anchoita*, fueron encontrados en cada campaña oceanográfica, lo que significa que esta especie se reproduce a lo largo de todo el año. La intensidad de su reproducción cambia de acuerdo con la época del año.

2) La frecuencia de huevos y larvas de anchoíta en comparación con otras especies de peces es muy elevada.

3) La gran densidad de huevos de anchoita, siquiera en comparacíon con la anchoveta peruana, indica la existancia de muy grandes recursos pesqueros de esta especie primeramente mencionada.

4) El área más importante de desove de la anchoíta se encuentra ubicada, más o menos, en frente del Río de La Plata. Sus limites se desplazan en dirección sureña o norteña de acuerdo con la época del año.

5) Existe una relación positiva entre la presencia de huevos de la anchoíta y la de otras especies de peces. En gran cantidad de los casos al mayor número de huevos de la anchoíta acompañaba mayor número de huevos de otras especies.

6) En la mayoría de los casos el grán número de huevos de anchoíta fue relacionado con pequeño número de los organismos zooplanctonicos. Se supone que los grandes cardúmenes de la anchoíta desovante puedena-fectar seriamente las poblaciones de zooplancton.

INTRODUCTION

Some years ago in the Institute of Marine Biology, Mar del Plata, Argentina investigations began on marine fish eggs and larvae. Considering the importance of these subjects, the studies were as much from the basic scientific point of view as for fish resources evaluation. A big impulse in the development of the eggs and larva surveys was the colaboration of this Institute with the Fishery Development Project which is carried out by the Argentinean Government with the co-operation of the United Development Program and the Food and Agriculture Organization (FAO). Thanks to the collaboration of the two institutions it was possible to undertake several oceanographic cruises which allowed numerous samples of plankton to be obtained at different seasons of the year and over a relatively wide area of the sea.

The first aim of these surveys was the determination of the distribution and density of anchovy (*Engraulis anchoita*) eggs and larvae and an estimation of the spawning intensity of this species. The results of these investigations were published earlier (Ciechomski, 1965, 1968a, 1968b, 1969). With the anchovy the eggs and larvae of other species of fish were also collected. Many of them could not be determined because of the lack of information on the taxonomy of marine fish eggs and larvae from this part of the ocean. Nevertheless, some observations could be made. Another fault of the investigations was the fact that the whole area of the surveys could not be covered entirely during every cruise.

In spite of these failures some general considerations are presented referring to the ichthyoplankton problems in the shelf waters of the Southwestern Atlantic Ocean off shore of a large part of Argentina, Uruguay, and the southern part of Brazil.

MATERIAL AND METHODS

At the present time the systematic egg and larva surveys covered three yearly cycles. Plankton samples were obtained during more than 20 oceanographic cruises and from many short trips close to the shore. In the present paper only the results of the most representative 9 cruises performed between August 1967 and June 1969, and of 6 short trips close to the shore (5–15 miles off shore) in the area around Mar del Plata performed in the most intensive reproduction season of the anchovy of 1967 and 1968 are presented. The results obtained in the remaining cruises and short trips were in some aspects more or less similar to those obtained in the cruises mentioned above, and for this reason are not presented in the present paper.

The areas covered by the cruises are shown in the Figure 1 and included a zone between the parallels 30° 44′ S and 47° 55′ S. Between approximately latitudes 30° 44′ S and 40° S the stations covered the entire shelf, reaching in some occasions waters some miles off the slope. Between the parallels 40° S and 47° 55′ S the area of the investigations included only a part of the shelf which is very extended in these latitudes (Fig. 1).

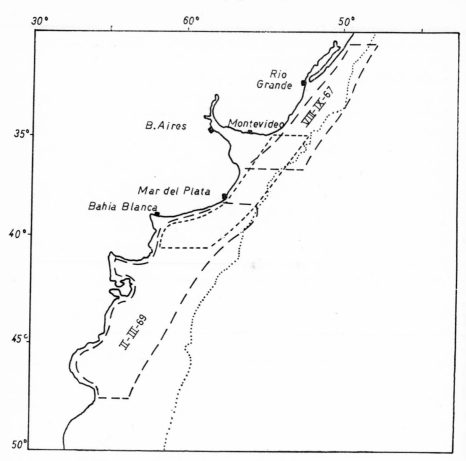

Figure 1 Areas covered by the cruises. Explanation in the text

Unfortunately the entire area was not investigated on each cruise. The most northern and the most southern areas indicated in the figure (long dashes) were investigated only once during the period of surveys. The intermediate area (short dashes), between approximately 35° S and 40° S, was almost covered in all cruises.

The quantitative plankton samples were collected with a Hensen net with a mesh of 325 microns. Vertical hauls from the bottom, or from 100 m in deeper waters, to 0 m were taken. The numbers of eggs and larvae were calculated as the number per square meter in the corresponding column of water. The average values per station which reflect the spawning intensity in the whole area, and average values per positive station which reflect the spawning intensity within limited boundaries of the spawning area were calculated. Larvae only in the youngest stages, less than 10 mm in length, were considered.

At every station data of temperature and salinity of the surface water were taken and bathythermograph observations from 0 m to the bottom were made.

RESULTS

The mean values of the density of fish eggs and larvae per square meter found at different seasons of the year during different cruises are shown in the Table 1.

The southern and northern limits of the area covered are pointed out. In column 6 and 10 the total number of all fish eggs and larvae respectively per square meter are given. From these total numbers the eggs of only two species, the anchovy *Engraulis anchoita* and the mackerel (caballa) *Scomber japonicus marplatensis*, are considered separately, and the percentage of their frequency in relation to the totals are given.

Anchovy eggs and larvae in plankton

Although, the spawning intensity of fish species generally, and particularly of the anchovy and the mackerel, varies according to the season of the year, fish eggs and larvae were found in the plankton at all seasons. This fact demonstrates that this species reproduces during the whole year.

The frequency of anchovy eggs and larvae in relation to all other species is in every case very high. The highest percentage (94.7 and 96.0) was found in the cruises during October and November, months when the reproduction of this species is most intensive. The similar phenomenon can be observed in the results obtained in the short trips around Mar del Plata during the same months. At some stations the number of anchovy eggs and larvae exceeded 15,000 and 5,500 per square meter respectively. But even when the spawning of the anchovy is least intensive, as in June or July, the percent values are always high and the predominance of eggs and larvae of this species was evident. There is no species of fish whose eggs and larvae could be compared in number with those of the anchovy. The only exception can be observed in the coastal stations of the cruises during February and March which touched the southern most limits of the area investigated (47° 55' S).

Table 1 Number of fish eggs and fish larvae collected

Date of the Cruise	North. and South. Limits	Range of Temp. °C		Number of fish eggs/m²	
		All Stat.	Posit. St.	per st.	per pos. st.
30/VIII–6/IX 67	30°44′–36°12′ (26)	8.8–19.7	10.4–13.1	137.9	211.0
31/X–12/XI 67	36°30′–41°00′ (50)	9.7–16.9	9.7–16.9	901.2	1047.9
20/I–8/II 68	37°00′–41°00′ (43)	12.2–21.5	12.2–21.5	96.6	112.3
18–24/IV 68	35°00′–38°00′ (30)	13.4–17.7	13.4–17.7	129.0	143.4
13–26/VII 68	34°30′–40°30′ (60)	5.9–13.7	10.2–13.7	52.2	116.0
16–27/X 68	36°00′–40°40′ (48)	12.2–15.8	12.5–15.8	1609.2	2032.6
27/XI–6/XII 68	36°18′–39°10′ (28)	16.7–23.8	16.7–23.8	649.3	649.3
22/II–23/III 69	39°19′–46°26′ (16) of shore area	13.0–18.5	14.8–18.5	439.0	548.7
22/II–23/III 69	38°44′–47°55′ coastal (56) area	13.2–20.6	13.2–20.6	38.8	63.9
14–20/VI 69	38°00′–40°11′ (23)	10.0–14.0	11.5–13.0	28.8	66.2
20–21/X 68	37°50′–38°10′ 5–15 mil. (30) off sh	12.5–14.0	12.5–14.0	1128.7	1128.7
23–24/XI 68	37°50′–38°10′ 5–15 mil. (30) off sh	16.7–17.8	16.7–17.8	6595.0	6595.0

(In brackets: number of stations).

Generally, in this season of the year, the anchovy shoals are spread over the shelf further off shore than in the spring or the beginning of the summer.

It is worth mentioning that the parallel 47° S is considered as the southern most limit of anchovy distribution, and the eggs and larvae of this species were never found in higher latitudes. The analysis of the investigations on the ichthyoplankton in the more southern zones of the ocean will not be treated in this paper. At the present time we have too little information on this matter.

We shall not enter into details of the distribution of the ichthyoplankton; we wanted only to draw attention to the large abundance and the evident predominance of anchovy eggs and larvae in the sea in the extensive investigated region and in every season of the year. Even in comparison with the abundance of the eggs of "anchoveta", *Engraulis ringens* in Peru and Chile (Brandhorst and Rojas, 1968: Rojas de Mendiola, 1964, and others) the density of the eggs per square meter of the Argentinean anchovy seems to be very high. Furthermore, the presence of its eggs in the plankton during

during different cruises (per 1 square meter)

Per cent anch. eg.	Per cent mack. eg.	Number of fish larvae/m²		Per cent anch. l.	Salinity	
		per st.	per pos. st.		All st.	Pos. St.
88.0	—	37.0	53.4	31.9	24.91–36.62	26.73–33.62
94.7	—	18.6	51.6	92.2	29.70–34.50	29.70–34.50
71.1	0.9	84.1	124.7	74.6	24.40–35.84	24.40–35.84
58.4	—	42.2	52.8	56.4	33.30–33.73	33.38–33.73
93.2	—	3.9	13.1	64.1	29.02–34.16	29.02–34.16
96.0	—	46.1	79.1	86.6	27.00–34.34	27.00–34.34
72.8	10.4	73.1	102.3	65.4	27.00–33.76	27.00–33.76
76.8	—	185.6	232.0	83.4	33.27–33.64	33.27–33.64
14.1	—	27.6	83.4	88.4	32.89–34.24	33.28–34.24
89.8	—	5.0	9.5	100.0		
94.6	—	12.0	12.0	92.0	33.17–33.38	33.17–33.38
95.7	0.6	1160.0	1160.0	96.5	33.20–33.48	33.20–33.48

the entire year would suggest the existence of very large fishery resource for this species.

As in the investigated area the anchovy eggs and larvae predominated at every season of the year, and as there is no species whose eggs and larvae could exceed in number those of the anchovy, it seems that lack of competition has a positive influence on the survival of the larvae of this species. The species whose eggs and larvae were found in relative abundance but only at a limited period of time was the "caballa", *Scomber japonicus marplatensis*. In Peru and Chile some competition might be expected on the part of the sardine, *Sardinops sagax* since in the California region the competitor of *Engraulis mordax* is the sardine, *Sardinops caerulea*.

This great abundance of Argentinean anchovy eggs and larvae, even when compared with the Peruvian anchoveta, is of greater interest since *Engraulis anchoita* (a zooplankton feeder) is situated on a higher trophic level than *Engraulis ringens* (a phytoplankton feeder). Nevertheless, it is worth mentioning that both species, as demonstrated by Ciechomski (1967) and further by Rojas de Mendiola *et al.* (1969), feed on zooplankton from their early stages until their larvae reach 45 mm in length.

Another fact should be mentioned is that there are two principal spawning areas of the anchovy, one in the north and one in the south. The more important seems to be the northern area, located more or less in front of the Río de La Plata. The limits of this area vary in a southern or northern direction, according to the season of the year. It could be related to some influence of waters of the mentioned river and of the vicinity of the Subtropical Convergence. For elucidation of this problem, however, a detailed analysis of hydrochemical conditions and primary production of this region should be made.

Co-occurrence of anchovy and other fish eggs

Analyzing the results obtained during these investigations an interesting observation can be made. In the majority of stations the eggs and larvae of other species of fish were found in the same samples where anchovy eggs and larvae occurred. In the majority of samples where no anchovy eggs were found, other fish eggs did not occur. In many cases larger numbers of other fish eggs were associated with larger numbers of anchovy eggs.

The problem of co-occurrence of anchovy (*Engraulis mordax*) and sardine larvae in the California Current region was considered by Ahlstrom (1967). He found that the number of anchovy larvae when taken in hauls containing sardine larvae were usually double the number found in hauls containing only those of the anchovy. Also, numbers of sardine larvae in samples that contained anchovy larvae were usually considerably higher than in those hauls where they occurred alone. But the frequency of co-occurrence of the eggs of both species was markedly lower than for the larvae. Analyzing this problem Ahlstrom (1967) came to the conclusion that the high frequency of co-occurrence of larvae of both species, even when the number of larvae per haul were large, did not indicate better conditions for survival.

Our observations led to the conclusion that the co-occurrence of eggs of these species reflect some specific hydrological conditions. These conditions can stimulate the spawning of the majority of species of fish in the season of their reproduction, or could act as an inhibitory factor.

The problem of an eventual competition for the food in areas of large concentrations of larvae seems to be of little importance. The relatively high levels of primary production in the shelf waters of the investigated area

(Fernandez, Lusquinos and Orlando, 1968), especially at the season of the most intensive reproduction of the anchovy and other species of fish, would appear to be sufficient to support an adequate zooplankton population which in turn could support the fish larvae population.

Anchovy eggs and zooplankton

Another interesting observation resulting from the obtained data is the fact that in the majority of cases the large numbers of anchovy eggs were related to small numbers of zooplankton organisms. Conversely, in the samples with high zooplankton biomass (especially copepoda) no anchovy eggs or only a small number were found. A similar phenomenon was observed in the case of the Peruvian anchovy (Flores, 1967; Guillen and Flores, 1967). The problem is discussed by Cushing, (1969) who gives alternative explanations: 1) the spawning anchoveta eat algae competing with the zooplankton, or 2) the anchoveta eat small zooplankton animals before they spawn.

In case of the Argentinean anchovy the explanation of this phenomenon would seem to be easier. This species is a typical zooplankton feeder and the large shoals of spawning anchovy could seriously affect the zooplankton populations. This problem should be investigated in more detail since the newly hatched larvae are feeding primarily on eggs, nauplii and young stages of copepods (Ciechomski, 1967).

Acknowledgements

The author wishes to express her gratitude to Miss M. Amalfi and Miss A. Valdetarro for the laborious work of counting eggs and larvae in the plankton samples. Also her gratitude goes to all the persons who collaborated in collecting of material and who put oceanographical data at her disposal.

References

AHLSTROM, E. H. (1967). Co-occurrences of sardine and anchovy larvae in the California Current Region off California and Baja California. *CalCOFI, Rep.* **11**, 117–135.

BRANDHORST, W. and ROJAS, O. (1968). Investigaciones sobre los recursos de la anchoveta (*Engraulis ringens*) y sus relaciones con las condiciones oceanográficas en agosto–octubre de 1964. *Inst. Fom. Pesqu., Publ.* **36**, 1–17.

CIECHOMSKI, J. D. de (1965). Observaciones sobre la reproducción, desarrollo embrionario y larval de la anchoíta argentina, *Engraulis anchoita. Bol. Inst. Biol. Mar.* **9**, 1–29.

CIECHOMSKI, J. D. de (1967). Investigations of food and feeding habits of larvae and juvenils of the Argentinian anchovy, *Engraulis anchoita. CalCOFI, Rep.* **11**, 72–81.

CIECHOMSKI, J. D. de (1968a). Distribución y abundancia de huevos y larvas de anchoíta en la region bonaerense y Norte Patagónico. (Agosto 1966–julio 1967). *Publ.* **4**, *Proy. Des. Pesqu., Ser. Inf. Téc.* 1–7.

CIECHOMSKI, J. D. de (1968b). Distribución estacional de huevos de la anchoíta (*Engraulis anchoita*) en el Atlantico Sud-Occidental. *CARPAS/4, D. Téc.* **31**, 1–8.

CIECHOMSKI, J. D. de (1969). Investigaciones sobre la distribución de huevos de anchoíta frente a las costas argentinas, uruguayas y sur de Brasil. Resultados de nueve campañas oceanográficas, agosto 1967–julio 1968. *Publ. 14, Proy. Des. Pesqu., Ser. Inf. Téc.* 1–10.

CIECHOMSKI, J. D. de and BOSCHI, E. E. (1968). Resultados de salidas costeras frente a Mar del Plato para el estudio de huevos y larvas de peces y crustaceos comerciales. Ano 1967. *Publ. 6, Proy. Des. Pesqu., Ser. Inf. Tec.* 1–8.

CUSHING, D. H. (1969). Upwelling and fish production. *FAO Fish. Tech. Pap.* **84**, 1–40.

FERNANDEZ, L. C., LUSQUIÑOS, A. and ORLANDO, A. (1968). El Servicio de Hidrografía Naval y el Proyecto de Desarrollo Pesquero en Argentina. *CARPAS/4. D. Téc.* **41**, 1–56.

FLORES, L. A. (1967). Informe preliminar del crucero 6611 de la primavera de 1966 (Cabo Blanco–Punta Coles). *Inf. Inst. Mar. Peru.* **17**, 1–16.

GUILLÍN, O. and FLORES, L. A. (1967). Informe preliminar del crucero 6702 del verano de 1967 (Cabo Blanco-Arica). *Inf. Inst. Mar. Peru.* **18**, 1–17.

ROJAS DE MENDIOLA, B. (1964). Abundancia de los huevos de anchoveta (*Engraulis ringens*) con relación a la temperatura de mar en la región de Chimbote. *Inst. Inv. Rec. Mar., Inf.* **25**, 1–24.

ROJAS DE MENDIOLA, B. *et al.* (1969). Contenido estomacal de anchoveta en cuatro areas de la costa Peruana. *Inf. Inst. Mar. Peru.* **27**, 1–30.

Physical factors in the production of tropical benthic marine algae

MAXWELL S. DOTY

Botany Department, University of Hawaii, Honolulu, Hawaii

Abstract

Sizes of the larger frondose algal standing crops are controlled on open tropical reef flats as much by the unpredictable storm turbulence as by regular seasonal factors. The smaller less ambiently frondose algae are less affected by storms and are more seasonal than the larger algae. The diffusion rate enhancement effects of water movement are major requirements for growing the larger benthic algae. Calcium sulfate "clod-card" dissolution rates provide measurements of this diffusion rate enhancement that correlate this hypothetical cause well with the standing crop sizes which are thought to be its results. Light intensity has been found to bring on aging such as may be related to phenomena like the shifts in carbon to nitrogen ratios that lead to flowering and death in determinate flowering plants. Tidal control of many fertility factors for benthic algal growth is recognized. Perhaps chief among them are control of water movement, its mixing or exchange, control of light through water depth, turbidity and control of predators. One might say that the influence of tides and other forms of water motion are major among the physical variables within the tropics in respect to fertility of the sea for benthic algae through its influence on other requirements.

Resumen

A dimensão dos "standing-stock" de algas benticas é controlada em recifes de coral abertos, tanto por fatôres como a turbulência, como por fatôres sazonais regulares. As algas menores com frondes menos desenvolvidas são menos afetadas por turbulencias e tormentas do que as maiores. Os efeitos da difusão do movimento de água são fatôres importantes no

crescimento das algas bênticas de maior tamanho. Razões de dissolução
de sulfato de calcio fornecem medidas da razão de difusão, as quais per-
mitem correlacionar essas causas com a dimensão do "standing-stock".

O contrôle pela maré de muitos fatôres de fertilidade, para o crescimento
de algas bênticas é reconhecido. Talvez os principais sejam o contrôle
de movimento de água, turbidez, contrôle de penetração de luz através
da água e contrôle de predadores. Pode supor-se que a maré é o principal
fator que influencia a fertilidade de algas bênticas em águas tropicais
através de sua influência em outros fatôres.

INTRODUCTION

Fertility of the sea for the algae is largely controlled by the peculiarities of
variation in the physical environment. The nature of algae, themselves,
is such that they gain and dispense materials by diffusion. Inorganic materials move into them and some of these materials become deposited in
compounds or forms peculiar to the species. This process is called primary
production. Sunlight is the energy source. Thus, the algae are shallow water
organisms. Their primary production in its various forms is the predominant
and most important of their processes; though they also carry on mineral
production and secondary production as well.

As an algal spore grows in size differentiation occurs and, according to
the species, the algae generally comes to assume one of three growth forms.
These three generalized forms are listed here as planktonic, microphytic
and macrophytic. They differ conspicuously, especially in their relationship
to the surrounding inorganic environment.

The planktonic forms are usually unicellular and microscopic. They
dominate the vast oceanic habitat where the concentrations of fertilizer
elements are very low. *Coccolithus*, *Ceratium* and *Halosphaera* are examples
of the larger forms and are more commonly in the open ocean. The small
open ocean forms are hardly known. *Olisthodiscus* and *Chrysochromulina*
are characteristic of the smaller and *Skeletonema* is characteristic of the
larger inshore forms. Such are phytoplankton.

"Microphytic" is a term to be used for the generally cylindrical filamentous kinds which are one-cell thick. They are commonly prostrate
against, near or buried in the substratum. Their environment is usually mud
or other solid material. Inorganic matter is more highly concentrated in
their environment than in the case of the other two forms. Many of the
blue-green algae would serve as examples. *Cladophora* and *Vaucheria* are
good examples of large microphytes. *Caulerpa* is a yet larger but more

typical example in that it would seem to be a heterotrichous elaboration of the basic prostrate tubular form. Here penetrating rhizoidal filaments enhance contact with the inorganic materials in the environment; perhaps compensating for the elaboration of erect fronds.

The macrophytic forms are multicellular and massive. They are the benthic algae characteristically found in moving water. They depend on attachment to hold them in place with the result (Ruttner, 1926) that the water moves past them and enhances diffusion. *Chondrus, Sargassum* and *Gracilaria* are three of the more widely known. Less well known of the algae in this category is the major builder of atolls and low islands in the Central Pacific, *Porolithon*, and the important carrageenan producer, *Eucheuma*. The macrophytic algae are the economic seaweeds.

The sea appears to be reciprocally fertile in respect to the phytoplankton, the micro- and the macrophytes. That is to say, a place ideally fertile for one of these forms is not ideal for the other forms. Ruttner (1926), Feldmann (1938) and others have observed that benthic populations are reduced as calm or stagnant water is reached. On the other hand, Chandler (1937) felt the opposite to be true in respect to phytoplankton. Little knowledge has been added since.

Though this discussion is restricted to the macrophytic algae, it should be noted that ultimately the primary production of all three forms becomes organic or inorganic detritus or enters the food chains. The entry into the food chains is either by grazing or as dissolved matter.

The inorganic productivity of marine algae is impressive. Calcium and carbon deposited as carbonate produce most of the consolidated reef structures in the Tropical Pacific. Algal carbonate contributes a majority of the sediments; sand and rubble accumulate both on the sea bottom as calcareous ooze and on the reefs as islands. Silica is deposited as siliceous ooze. The vast and commercially important diatomite deposits are algal also. The fixing of nitrogen and perhaps the aminization process in the sea is algal as well as the entry of phosphorus into biological systems. However, we will bypass these inorganic production phenomena.

The principal subject of this paper is the mechanics of the sea's apparent fertility in respect to macrophytic or benthic algal organic production. Apparent fertility is what we presume from observations of standing crops, growth rates or other single observations. Something akin to Heizenberg's theory in physics applies here. Real or absolute fertility is obscured as soon as we control one factor without controlling all others; so as a result no absolute measurements of fertility can be made. Yet isolated as major mechanical factors controlling fertility, we use wind-generated waves, tides, diffusion and sunlight. Other factors conventionally considered, such as temperature and fertilizer content, are secondary, i.e. their variations are largely controlled by these major or primary factors.

STORM TURBULENCE CONTROL OF FERTILITY

The standing crop of benthic algae averages 2.3 kg/m^2 on one Hawaiian reef surface studied for several years. Ryther and Menzel (1960) concluded weather was a major factor in the lower early and higher late winter or early spring productivity of plankton in waters such as those off Bermuda where Bernatowicz (1952) found a similar seasonality in the benthic algae. At the same latitudes in Japan, Hayashida and Sakurai (1969) have also reported seasonal variation in size of the standing crop. These latter authors reported a maximum (2.3 kg/m^2) in April and a minimum crop (1.7kg/m^2) in August. No quantitative data are available for areas nearer the equator. This is not to deny the value of the recent detailed publication of Earle (1969) on the ecology of the Phaeophyta of the northeastern Caribbean nor that of Svedelius (1906a and b) for Ceylon nor the various studies of F. Boergesen (e.g. 1911) and G. W. Lawson, respectively for the Virgin Islands and Ghana but which, with the exception a brief abstract (Lawson, 1959), do not even imply that studies were done in respect to mass per square meter.

The Hawaiian area, at 21 deg. N. Lat., (Fig. 1) is along the beach at Waikiki in Honolulu. It has been sampled by a controlled random sampling method at intervals throughout the year with over 100 samples being taken each time to represent about 0.2 per cent of the 33,600 square meter area. The 2.3 kg/m^2 value is for wet weight of harvestable, i.e. frondose, macrophytes. The Sulu area (Fig. 1) has likewise been studied in somewhat the same way, with the mean of 0.1 kg/m^2 being found.

Such standing crops of benthic algae vary with time. In Hawaii at Waikiki (Fig. 3) the variation is between 1.3 and 3.1 kg per square meter. The principal causes of such standing crop variation in temperate regions are combinations of seasonal events such as weather, light and temperature. In the tropics light and temperature are far more nearly constant. In Hawaii the sea temperature can be expected to vary as much as about 4 degrees during the year. Light may vary from its lowest values at this site in February

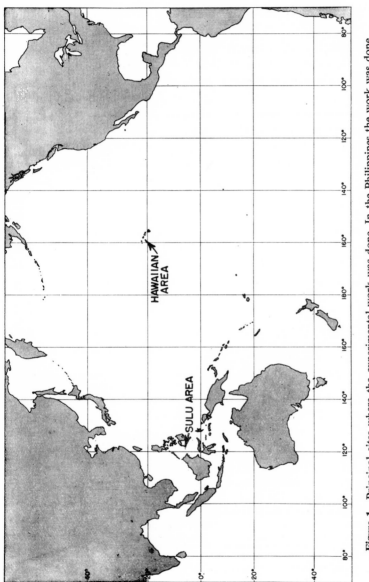

Figure 1 Principal sites where the experimental work was done. In the Philippines the work was done largely at Zamboanga (near 7 Deg. N. Lat. and 122 Deg. E. Long.) and in the Sulu Archipelago in co-operation with the Philippine Fisheries Commission. In Hawaii the work was done at the University of Hawaii beach laboratory at Waikiki in Honolulu (near 21 Deg. N. Lat. and 157 Deg. W. Long.)

to twice as much in August, the sunniest month. Thus, seasonality is expected and, indeed, in working with the data one can find (Fig. 2) various seasonal patterns of variation in the relative abundance of the different individual species.

The variation in total mass per unit area (Fig. 2) is not found to vary seasonally in a simple manner at Waikiki. An explanation for the irregularity was sought in the weather which is more irregular than seasonal. Among other things, the wave (i.e. swell) height records for the leeward shores of the island of Hawaii 150 miles away, but exposed as is the study area, were summarized along with those from Makapuu, a more windward site some

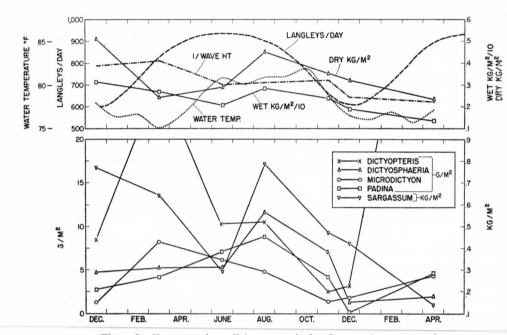

Figure 2 *Upper panel:* sunlight energy in langleys at the upper surface of the atmosphere at 20 degrees North Latitude; the reciprocal of wave height index, 1/h, for the 3rd week previous to each of the harvests, which are shown as 0.1 times the kg wet weight per square meter (e.g. 1.33 for Apr. 68) and as kg dry weight per square meter. *Lower panel:* standing crop as kg or grams wet weight per square meter for *Sargassum echino-carpum* and dry weight for the other species which are *Dictyopteris plagio-gramma, Dictyosphaeria cavernosa, Microdictyon setchellianum* and *Padina japonica.* The March, 1967, dry weights have been adjusted from the wet weights by use of the mean wet/dry wet ratio for the species

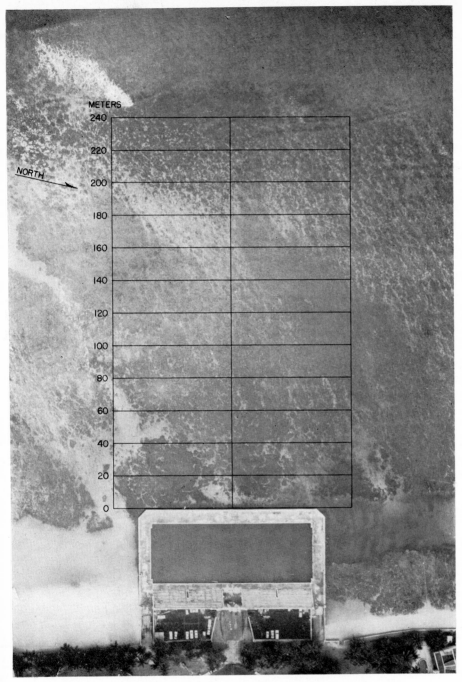

Figure 3 The study area at Waikiki, Honolulu, Hawaii, between the War Memorial Natatorium and the reef edge. This area was divided into 20 meter-wide zones, each 140 meters long an each divided again into north (at the right) and south halves

10 miles east of the study area. This was done by making a time-course plot of the wave heights for the seven weeks during and before each harvest period. The area under each week's curve was determined and recorded (Table 1) as a wave height index value.

In studying the correlations between wave height index and standing crop size, little correlation ($r = 0.15$) is found between the crop and the leeward wave heights. On the other hand, the weekly values similarly obtained from the Makapuu site (Table 1) correlate well with standing crop for some prior weeks. However, there is little correlation with the sixth, fifth or fourth week's values and none with the first and the contemporary week's values. The correlation improves the second week previous and if it (Table 1, 0.73) were -0.75, it would be significant at the five per cent level. Then, a correlation of -0.89, significant at the one per cent level, was obtained for the third week prior to the measurement period. The regression is high for the second and third week's values and is much less high for the fourth and fifth weeks.

Both the high negative regression and the high correlation between standing crop and wave height, a correlation improving toward the third week previous, lead us to conclude that sea turbulence is a major controlling factor of size variation in the benthic algal standing crop. Thus, in Hawaii

Table 1 Sea wave height index values in respect to mean algal standing crop harvested at Waikiki in Hawaii. The index values are the daily average area found under a curve of swell height recorded at Makapuu plotted as a function of time. These units are arbitrary

Date of harvest '66–'68	Sea tubulence							Algal $\bar{x}$ wet wt kg/m²
	Six weeks prior	Five weeks prior	Four weeks prior	Three weeks prior	Two weeks prior	One week prior	Contem-porary week	
Dec.	2.73	2.63	4.77	2.57	2.50	4.38	3.36	3.12
Mar.	5.00	3.15	3.62	2.43	1.77	2.31	3.88	2.67
Jun.	3.39	2.46	1.75	3.30	3.96	1.96	2.71	2.06
Aug.	3.81	3.85	3.19	3.18	3.71	3.21	2.73	2.88
Nov.	2.96	2.50	3.00	3.14	3.58	4.43	5.93	2.34
Dec.	2.46	5.36	6.50	4.09	4.37	2.73	2.28	1.90
Apr.	3.04	5.31	5.68	4.65	4.39	4.50	3.50	1.33
b	0.23	−0.30	−0.13	−0.70	−0.46	−0.02	0.04	
r	0.32	−0.62	−0.34	−0.89	−0.73	−0.03	0.08	

where there is no strong seasonal storminess but rather random occurrences of storms or high waves, random weather often dominates the standing crop size and seasonality is obscured. In monsoonal areas (Svedelius, 1906b) or where there is a regular season (Lawson, 1957a) of rough water, i.e. turbulence, a seasonal effect on the standing crop can be expected such as reported by Svedelius (1906b), Lawson (1957b) and others.

Other aspects of variation in standing crop can be recognized, such as seasonal changes; though they are perhaps less dramatic. The dry standing crop weight (Fig. 2) of such smaller macrophytes as *Padina japonica* ($r = 0.81$) and *Microdictyon setchellianum* ($r = 0.82$) correlates much more closely with the seasonally changing values for light (List, 1966) than with the wave heights and this light phenomenon is much less important for the taller, more frondose, *Dictyopteris plagiogramma* ($r = 0.55$) and *Sargassum echinocarpum* ($r = 0.39$). Furthermore, the regression is positive for *Padina* and *Microdictyon* and, at least negative for *Sargassum*.

In the case of *S. echinocarpum* the correlation ($r = 0.74$) of standing crop dry weight with wave height would seem to indicate wave height is more important in controlling standing crop than light. This is not true in the case of the other two species of *Sargassum*. This difference between the species would seem to reflect a differential response that should be relatable to form or habitat ecology. As an example, the taller species, e.g. *Dictyopteris* and *Sargassum* in general, appear to be more susceptible to wave action than are the shorter less-frondose species. As an example of habitat difference, *S. polyphyllum* tends to grow more deeply in the water and bloom during the more brightly lit months of the year than *S. echinocarpum*.

Projecting this idea, one is led to expect the lowest story, the crustose forms, to show the least variation with storm conditions. However, mere low stature is not everything. Variations in the standing crop of the 3 to 5 cm high, rigid *Dictyosphaeria cavernosa* are the least closely correlated ($r = 0.32$) with seasonal light variation of all those we have studied and is, on the other hand, one of the species more closely correlated ($r = 0.51$ for dry weight and 0.59 for wet weight) with wave height. It is noted that *Sargassum echinocarpum* and *D. cavernosa* are often the major components of the wrack cast onto the beaches after storms.

In this work the wet and dry weights do not correlate as well as one might hope. This is thought to be related to changes in the wet to dry weight ratios obtained during the year in the different species. For example, Haug and

Jensen (1954) have shown there may be a variation of over 15 per cent in the ash in *Alaria esculenta* during the year.

TIDAL CONTROL OF FERTILITY

In the tropics the tidal rise and fall of the sea's surface is perhaps most important for its control of water movement. On the other hand, in the Temperate Zone the tides are more important for the way (Doty, 1946) they periodically remove the water from the intertidal region. Japanese seaweed production, which is in the Temperate Zone, depends to a large extent on intertidal exposure of the cultures. In the tropics there is very little in the way of a standing stock of algae in the intertidal region. The tides are, thus, of most importance as they induce currents and turbulence on tropical reef flats and often the communities are distributed in reference to these phenomena. The sea is generally most fertile for tropical macrophytes on the tops of coral reefs.

Reef flat surfaces are in general at the levels of the lowest tides. How they come to be at this level so often, has almost as many explanations as there are for the origin of reefs themselves. Coral reefs which produce reef flats occur in the Western Pacific and Atlantic as far north as Kagoshima, Japan, and as far south (McMichael, 1966) as Lord Howe Island, and in the Bermuda area. These are places 31 to 32 degrees poleward from the equator. When separated from shore by deep water, they are often referred to as patch reefs, or they may appear as atolls which often bear islands of coral debris which support terrestrial life.

Calcareous organisms that can withstand the rain, the high salinities or desiccation, the intense insolation and the heat resulting from air exposure at low tide are few. Some of the melobesioid coralline algae are perhaps the most notable. They can withstand these effects of tidal exposure and the wave force at the seaward edges of reefs best. *Porolithon onkodes*, for example is the major reef-producing organism and most conspicuous living thing on Central Pacific atoll sea reefs. However, having the appearance of pink cement pavement, it is often overlooked.

Tidal exposure to air temperatures, insolation and desiccation is often critical. Many experiments (e.g. Biebl, 1962) have led to an understanding of the role of tidal exposure through its regulation of temperature, insolation and desiccation in restricting some algae to subtidal regions. In plankton experimental situations (Jitts *et al.*, 1964), insolation above optimal level may induce injury from heat accumulation in algal cells when it is not dissipated by cyclosis or motility. The barrenness of tropical intertidal regions is thought (Krishnamurthy, 1967) to be enhanced by high temperatures. In the case of the very successful culture practices for *Porphyra* and *Monostroma*, intertidal exposure is advantageous; since epiphytes and predators are reduced. Only a few tropical macrophytes grow higher than *Eucheuma* intertidally, but among them are *Gracilaria eucheumioides, G. (Corallopsis) salicornia* and some of the very small gelidioid algae. Algae such as *Eucheuma* are damaged by air exposure. Nevertheless, most of the *Eucheuma* crop comes from vertical ranges of height where the upper level is exposed to the air at a maximum of one or two hours.

Tides control entry of terrestrial ground water or water from the freshwater lenses of islands. There is a large literature on tidal control of estuary flushing and the entry of this water into the sea. This water is generally higher, but may be lower, in nitrate and phosphate and the ratio of these elements to each other is usually not that of sea water. If it comes onto the reef from a freshwater lens or from ground water, e.g. from springs at the shore, this water is usually much richer in fertilizer than the sea water with which it mixes. This water is probably enriched, like that in the disphotic zone of the sea, by downward movement of fertilizer from the soil above. The extremes to which tides permit this fertilizer-rich fresher water to move out onto reef flats have been related (Hiatt, 1957) to the horizontal distributions of reef organisms. River and estuary water that have been in the light for a long time have hydrophytes which remove the fertilizer so that, like the Amazon*, they may be lower in fertilizer than the sea water. Fresh water mixing with sea water in the light is influenced, depending upon circumstances, by turbulent tides induced near non-reef bordered shores or by height regulations of water running across reef flats.

Data gathered from stations within a few meters of shore in Kealakekua Bay, Hawaii, can be interpreted as showing (Table 2) differential uptake of fertilizer elements from such enriched water. In the case of the stations in Table 4 it is seen that the usual sea water nitrogen to phosphorus ratio of 15 (Cooper, 1938) is approached as the superficial lighter ground water is mixed in; though, as usual in such inshore waters, the ratio at these stations is variable over short distances. The mixing here appears to be with the

---

* See paper by E. D. Goldberg in this same volume.

Table 2 The content of water samples from stations along the shores of Kealakekua Bay, Hawaii

Station number	Depth in meters	Salinity $^0/_{00}$	NO_3—N μg A/l	PO_4—P μg A/l	N/P ratio
3	0	17.3	24.7	1.43	17.3
	1	34.0	2.5	0.21	11.8
5	0	24.7	17.2	1.27	13.5
	1	25.9	3.8	0.31	12.3
8	0	21.0	3.2	0.46	6.9
	1	36.8	2.5	0.22	11.4

water directly beneath the lighter overlying water. At Station 3 it is obvious that in diluting the runoff water with sea water, the salinity increases from 17.3 to 34.0. Also, the ratio of nitrate to phosphorus is higher than Cooper's or the ratio in the less saline water (17.3) would not approach the 11.8 ratio shown. At station 5, in addition to minor tidal mixing, the use of nitrate and phosphate in amounts about proportionate to their presence would seem to be the major process in changing the water from that at the surface to that just below. At Station 8 a case can be made for more rapid consumption of the phosphate relative to the common N/P ratios of 16–20 to 1, proposed by such people as A. C. Redfield and by R. H. Fleming.

Since the incoming water is mixed with sea water in Kealakekua Bay near where it comes in, differential ratios of use would seem to imply different users. It also seems probable that algal macrophytes, like *Ulva*, which are notoriously associated with incoming water play special roles both in lowering the fertilizer content and in shifting the nitrogen and phosphorus ratios. The currents within this bay reverse with the tides and do not, for the most part, carry such brackish water out of the bay. From the point of view of the qualities then, one can say tides are an influence on these users' community distribution and the fertility of the sea for them. Perhaps, also, the somewhat reciprocal distribution of riparian inflow and reef flat development, usually attributed to salinity, are related to these phenomena as well.

Alternately turbulent and calm, as well as turbid and clear conditions are provided by the tides as they rise and fall controlling the water depths over reef flats. Thus, tides having an influence on light and the enhancement of diffusion are of further significance in fertility of the sea for algal macro-phytes. These situations are taken up below.

Tides influence grazers strongly. Some coral fishes that may feed on *Eucheuma* move inshore at high tide only as far as the extreme lowest level of the tides. The aggressiveness of carnivores during rising tides and their relatively benign nature during falling tides is a well known factor tropi-cal biologists use in timing their ecological work on reefs. The sea urchins, *Diadema* and *Echinothrix*, are found clustered just below the extreme lowest tide levels. The lowest tidal limits of *Eucheuma* and most other macro-phytes coincide with these upward tidal limits of the sea urchins. Sea urchins are known elsewhere (e.g. North, 1964) as principal grazers reducing algal beds, and at least *Eucheuma* would seem to be pinned between the urchins below and an hour of air exposure above. Thus, a higher tide range would provide a larger biospace for production of this and other such genera.

DIFFUSION CONTROL OF FERTILITY

Macrophytes are characteristically attached to solid substrata in moving wa-ter. Insofar as the sea's fertility is concerned, it is a commonly made observa-tion that these benthic algae are large on the seaward ends of promontories and smaller as the shores of the promontory are followed toward calm water. Also their standing crops are smaller and the complement of species changes along such a gradient.

While water movement and especially the extremes of "force" (Jones and Demetropoulus, 1968) and "pressure" created by waves have been measured, they are usually only qualitatively correlated. Feldmann (1938) reviewed the rather extensive qualitative literature and concluded that in marine communities it is such mechanical action and regulation of tempera-ture through mixing that are of major importance. He also concluded that the effects of gas content and "purification" are negligible. Ruttner earlier (1926) had postulated the idea that the enhancement of diffusion by the movement of water increases the availability of the substances dissolved in it. Parenthetically, it likewise accelerates diffusion of substances from the organismsas well.

Experimentally, it has been found by Whitford (e.g. 1960) that benthic algal growth rates are enhanced by increasing water movement, an effect thought to be due to the steepening of gradients around the thalli. Ghelardi and North (1958) thought that marine animals and *Macrocystis pyrifera*

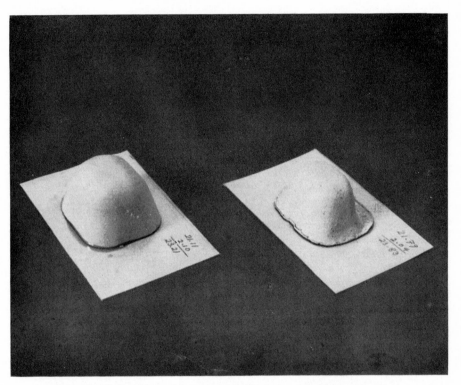

Figure 4 Clod-cards used in measuring the degree to which water movement enhances diffusion. The figures used in obtaining the initial dry
weigths are shown in the upper right corners of each clod-card. The clod-
card at the right was on the reef for one day in a relatively turbulent situation
and has lost about half of its original weight

are distributed, in part, as a result of the upwelling current velocities found of 45 cm per minute. Form has been related (Ott, 1967) to water currents and turbulence in the case of *Cystoseira*. Our own field observations on different *Caulerpa* species lead us to the hypothesis that diameter of the thallus is related to the degree of water movement. However, in none of these cases has the water movement and its degree of enhancement of diffusion been measured, aside from the applying of many devices (Welch, 1948; Gessner, 1950) to determining current velocity.

It seems that fertility can be recognized here as a function of diffusion in respect to distribution, community composition, growth rates or the other phenomena mentioned above. However, the lack of methods for measuring the diffusion-enhancing effects of water movement has been a barrier to testing this hypothesis. The methods must provide many simultaneous quantitative evaluations of water movements beyond the margins of a community and over the general area in which the community is found. Also, the method must provide a measurement of the combined effects of water movement in any direction regardless of whether the cause is tides, waves, currents or turbulence. Since over any tidal cycle in littoral regions there is a great deal of variation in this effect, the measurement devices must integrate the instantaneous values for whatever is chosen as the measurement period and provide a means for evaluating diffusion alone. Measurements of maximum wave force and single measurements of current velocity will not do. If many simultaneous measurements are to be made, the devices must be inexpensive, require little maintenance and be useable under extremely primitive field conditions.

In 1964 it was found that the dissolution of calcium sulfate from clods of Plaster of Paris can be used to estimate such diffusion enhancement. Following the suggestions made by different colleagues, over 40 other ideas for measuring this phenomenon have been tried as well. After this manuscript had been prepared, the author's attention was drawn to the publication of Bent J. Muus (A field method for measuring "Exposure" by means of plaster balls ... *Sarsia* **34**: 61–67), who used calcium sulfate for similar purposes and encountered many of the same problems in obtaining some of the same types of result.

Using calcium sulfate as clods cemented onto plastic cards (Fig. 4) and calibrated for their dissolution rates in still water, measurements yielding an index value for the enhancement of diffusion can be obtained. This index value is the weight loss in still water (the calibration value) divided into the weight loss for the clod-card preparation exposed at the chosen places for one tidal period. This clod-card method has been applied variously during the past few years in Hawaii, in Micronesia and in the Philippines. While there are problems, it does appear the method can be used in describing and comparing habitats.

As an example, index values were measured down the middle of the Hawaiian study area (Fig. 3) during the experimental harvesting of the standing crops discussed above in reference to storm control. A good correlation (Fig. 5) with standing crop size is usually obtained. In view of the few weeks' lag between the cause or control of standing crop size by storms and the measurement period, and the fact that simultaneous clod-card and standing crop measurements are being considered, it is not surprising that 0.70 (significant to the 1 per cent level) is the highest correlation found (Fig. 5) for this relationship to date.

The major technical problem found in applying this method involves what appears to be non-normal distribution of dissolution rates. This is revealed by about one out of five clod-cards giving a wild value when all are exposed under uniform conditions. This is handled by exposing the

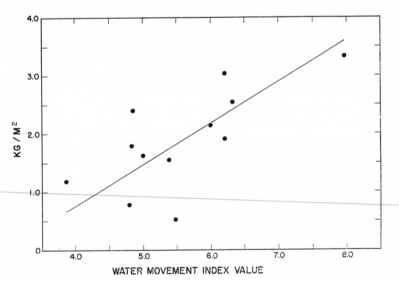

Figure 5   The standing crop plotted as a function of the clod-card measurements made December 21–22, 1967, along the center line of the study area at Waikiki. The regression slope is 0.87 and the correlation of 0.70 is significant at about the one per cent level

clod-cards in pairs and, with reasoned bias, discarding those values that appear to be damaged or otherwise non-conformative.

A solitary measurement period may coincide with some unusual circumstance, such as wind direction or force. An analysis (Table 3) of the results from the four measurements made along the center (Fig. 3) of the Waikiki study area during four separate tidal cycles can be taken here as an example. The correlations and regression slopes shown are generally significant. It would seem this standing crop and clod-card measurement period, itself, was (Table 1) a week of very "light" weather. However, the highest wave height indices were obtained (Table 1) for the weeks previous to their particular harvest and, until 2 weeks prior to this experimental harvest, there had been high waves resulting in one of the lowest standing crops measured. If results from one measurement are used, by chance one can get results which (Table 3, data for Dec. 28–29 in comparison to the rest) are quite in contrast to those more frequently obtained. For example, a negative regression and low correlation might be found on one day (Dec. 28–29) while the values for other times studied show reliable positive regressions and higher correlations. However, it is rather well indicated that the hypothetical relationship is sound and that faster growth is obtained where the water is measured by this method to be more turbulent.

As turbulence increases along a transect, the fertility of sites for the different species will generally increase. Some calm-water, small, finely-branched forms such as *Gracilaria confervoides* or *Hypnea valentiae* are not physically strong enough to withstand the higher turbulences and disappear in the most turbulent regions. Alternatively the explanation for this may be that under conditions of high turbulence, such thalli are mechanically damaged by outward diffusion exceeding rates of production. Conversely, in the lower diffusion gradients of calm water, massive macrophytes such as *Macrocystis*, *Lessonia* and *Postelsia* do not persist. In the case of the *Eucheuma* species, which grow near the edges of Pacific reefs, our experiments show the thalli grow faster in more turbulent water. In some cases turbulence may have its effects through control of predators or other algal species which would be competitors under different conditions of water movement. Thus, between periods of crop destruction, the standing crop grows the more quickly and, as a result, the standing crop measurements become larger where the water moves more.

On reef flats the distribution patterns often lead one to suspect water movement as the cause of community limits. While in such circumstances

Table 3 Clod-card measurement of diffusion enhancement compared to wet weight in 12 zones spread across the reef at Waikiki in Honolulu, Hawaii, in December, 1967. The clod-card values are mean loss of weight for clod-card pairs, in terms of grams of calcium sulfate lost over a 24-hour exposure period. For 10 degrees of freedom the 5% confidence level for correlation ($= r$) is 0.576; for 1%, 0.708. The mean regression ($= b$) and correlation values were calculated directly from the original data

Meters to zone center from shore	Mean wet wt of algae as kg/m^2	Clod-card weight loss for 24-hours exposure on the dates given				Mean clod-card weight loss
		21–22	22–23	28–29	29–30	
10	1.176	3.87	5.46	5.36	3.06	4.44
30	0.521	5.48	5.79	4.43	4.86	5.41
50	1.623	5.00	5.68	3.82	4.22	4.68
70	2.395	4.84	7.20	3.93	6.56	5.63
90	1.787	4.82	6.41	5.03	5.66	5.48
110	1.539	5.38	5.16	5.20	4.89	5.16
130	2.144	5.99	4.44	3.56	5.02	4.75
150	1.917	6.20	7.20	4.83	5.62	5.96
170	0.784	4.78	4.57	5.96	6.32	5.41
190	2.537	6.32	8.16	5.45	6.37	6.58
210	3.041	6.21	6.19	4.46	5.66	5.63
230	3.338	7.97	9.75	4.57	7.27	7.39
b		0.87	1.23	−0.27	0.75	0.60
r		0.70*	0.68	−0.31	0.56	0.62

* This value was 0.70322 before rounding.

tidal control of water movement is a dominating factor, tidal control of vertical distribution is minor. On reef patches isolated from shore by deep water, the tidal control of salinities postulated by Hiatt (1957) cannot be impugned.

Eucheuma striatum and its relatives seem to thrive on reef areas that have been called (Doty and Morrison, 1954) excurrent, i.e. reef surfaces over which during a tidal cycle drainage from the reef exceeds the flooding onto it. This could be an effect of temperature or ectocrines; as other species and genera peculiar to incurrent areas, i.e. areas over which during a tidal cycle flooding onto the reef exceeds drainage from it, such as *Porolithon* are reduced or replaced here.

In attempting to farm benthic algae, water motion to enhance diffusion is one of the most important fertility factors to be provided. It is especially necessary for macrophytes such as *Eucheuma*, whose massive thalli may be a centimeter or two thick and of hard brittle gel. Moving water requires power so costly that providing movement artificially cannot be considered. Obtaining rapid inward diffusion of fertilizer or other inorganic materials could be obtained by increasing the concentrations of these substances. In terrestrial farming, fertilizer can be added to increase the uptake rate of the crop for the fertilizer stays near the plant for some time. In marine open culture agronomy, fertilizer when added will quickly be diluted and carried away by the currents or utilized by weeds. Fortunately, it seems possible to develop circumstances on reef flats such that tidal ebb and flow will steepen gradients sufficiently that good economic growth rates will obtain.

LIGHT CONTROL OF FERTILITY

It is general knowledge that for algae in nature, light is often limiting and that, at least in phytoplankton, mature populations become narrowly stratified (Ryther and Hulburt, 1960) in reference to light intensity. Likewise, their chemical content (Yentsch and Vaccaro, 1958) may be altered. Yet, little is known of such non-photosynthetic light responses of the macrophytes.

It has been learned by bringing *Eucheuma striatum* and *E. spinosum* into successively shallower water that the brighter the light the faster the thalli grow as long as they are below the surface. These faster growth rates are about double those found in the natural habitats. Thus, a series of experiments has been carried out leading toward farming these seaweeds submerged in the top 15 cm of water on rafts at Zamboanga in the Philippines. These cultures are started from wild thalli.

Some thalli grow about twice as fast as others at the same light intensities. It is desirable to produce and maintain "seed" from such faster growing thalli, and, in order to reduce costs, keep reseeding to a minimum.

Under the bright-light raft conditions, initially the growth rates of the thalli increase (Fig. 6) and then decrease. The thalli weaken and the losses among replicates generally become so severe as to terminate the experiments when the growth rates drop lower than those obtained in nature. This phenomenon has become referred to as the "aging" effect.

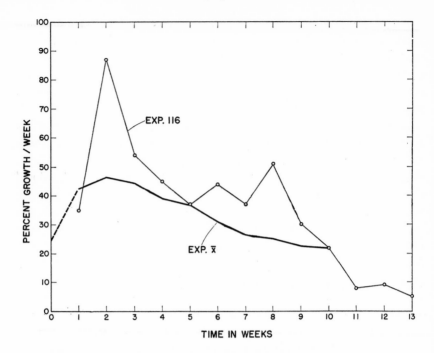

Figure 6 Variation in growth rate of *Eucheuma striatum* when held in the upper 15 cm of water at Zamboanga as a function of time lapsed from the start. The line "Exp. x̄" represents the mean values for the six control rate experiments started or being carried on during the same time period as "Exp. 116" for which only the brighter light values are shown. The curve for the mean values is damped strongly by variations in the week when the sub-initial maxima and later maxima occurred. The generally expected growth rate at the site in nature from which the thalli came is used as the zero time value

There is no correlation between these light phenomena or the aging effect and season and no *a priori* reason known for such Rhodophyta to be determinate in their growth. Thus, as a cause, it was postulated there is a difference between the water at the experimental and natural sites. To test this hypothesis, halves of aged thalli were transferred back to the site from which they had originally come. The halves there grew at essentially the same rates as those left on the floats. When ultimately some of these same thalli were brought a second time to Zamboanga (Fig. 1, "Sulu area") and placed on the floats, again the spurt in growth rate was obtained which we attribute to the brighter light. However, these thalli died soon. If the water at the Zamboanga sites was the cause of the aging, the return to natural site water should have cured it if the life span is not determinate. Thus, it was decided that water character alone was not the cause of this aging effect and that the life span of the thalli might have become determined by their exposure on the floats.

In terrestrial annual plants with determinate inflorescences, increasing light intensities may lead to flowering and death. In some cases such flowering follows increases in the ratio of carbon to nitrogen in the tissues. Thus, it was thought there might be a similar set of events in the experimental situation where the aging of Figure 6 appears as apposed to what happens in nature where no evidence of aging has been seen.

To test this hypothesis that the aging is due to changing the carbon to nitrogen ratios following exposure to the brighter light, halves of the same thalli were grown for 16 weeks on floats exposed to full sunlight while others were covered with nylon netting reducing the light about 50 per cent. Kjeldahl amino-nitrogen analyses in terms of per cent dry weight were made from sampling. As in previous experiments, following the initial high growth rates, lower rates followed (Fig. 6) and as individual thalli dropped in growth rate below about 20% per week, they were lost for one reason or another. So few thalli remained viable that none of the results is reliable insofar as they apply to the last part of the experiment when the light and shade conditions were reversed after the tenth week.

Three observations can be gathered from scanning the data (Fig. 7) from this experiment and its subsequent replications. First, there is a steady decline in nitrogen content for the first eleven weeks in both the brighter and dimmer light thalli. Second, the nitrogen content of the bright light examples was (Fig. 7) immediately and throughout the reliable first half of the experiment lower than that of the dim light examples. Third, while the initial growth rates of the bright light examples were higher, their *total* growth for the whole experiment was about the same as that of the dim-light halves of the same thalli.

A statistical examination (Table 4) shows similar degrees of dependence of growth on nitrogen content for the lot initially in the full sun (lot "A") and terminally in the full sun (lot "B"). The growth for both lots showed more independence from nitrogen content during their reciprocal periods in the shade.

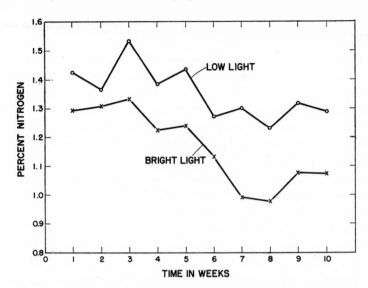

Figure 7 Variation in the nitrogen content of *Eucheuma striatum* thalli as a function of time in sets of thallus halves reared in the upper 15 cm of water at Zamboanga, Philippines, with the light striking one set reduced 50 per cent

Table 4 The regression of growth rate on nitrogen content of *Eucheuma striatum*. Both "A" and "B" were grown under the same conditions except that initially "B" was under a nylon mesh shade reducing the light about 50 per cent and "A" was without such shading. After 10 weeks the conditions were reversed on the two lots.

Expt'l lot	Sunlight conditions	
	Non-shaded	Shaded
A	56.5	40.0
B	48.5	9.8

The reversal of light conditions on the two thalli after ten weeks, if anything, merely postponed the decline and death of some of the thalli for the remaining weeks. However, in those remaining from the lot initially in bright light, the nitrogen content after the 12th week rose to a new level, though the thalli continued to decline in all respects as did those in the reciprocal lot. From these data it would seem possible the thalli under either light condition merely outran their nitrogen or other raw product supply.

In respect to light as a fertility factor it would seem certain that bright light may enhance the aging of thalli through inducing an imbalance between the rate of supply and use of some essential not affecting dry weight increase directly. There are, in this aging phenomenon, some similarities to shifts in the carbon to nitrogen ratios found in higher plants when they are shifted from dim to bright light. However, dipping thalli for a few hours in various concentrations of nitrate in sea water has not relieved this aging effect. Therefore, it may not be nitrogen itself that is responsible but some other substance. Possibly it is not even the unnaturally bright light, *per se*, but its heating effect or ultraviolet radiation. At any rate, in view of the lack of other means of producing selected "seed" under controlled conditions, this aging must be overcome if the potential fertility of the sea in respect to *Eucheuma* as a crop is to be realized.

In conclusion, in the tropics completely artificial benthic algal farming would seem to be technically feasible by designing the farms in respect to the physical factors so as to provide optimal water movement and optimal light. The vast reef flats would seem ideal for this purpose if protection from storms can be obtained. Subbaramaiah and Krishnamurthy (1967) felt that some artificial circumstances could be provided for indefinite growth of *Ulva, Sargassum, Cystoseira, Gelidiella* and *Gracilaria. Eucheuma*, which merely lives for months in bowls of still sea water, can be grown rapidly on rafts holding the algae in bright light. Here the small wavelets passing through the water in which the raft holds the algae near the surface provide the diffusion gradient enhancement needed. Selecting or manipulating the play of the tides is the cheapest way to obtain the exchange of water needed to provide water renewal and the diffusion rates which seem to be the major cost requirement for obtaining the long life and rapid growth needed for successful benthic algal agronomies.

Acknowledgement

It is a pleasure to acknowledge the financial support of the U. S. Atomic Energy Commission, the U.S. National Science Foundation's Sea Grant Program, Marine Colloids, Inc., and the U.S. National Institutes of Health, which agencies have made this work possible through various contracts and grants to the University of Hawaii. The diligence of Mr. Ernest Loveland in supervising the cooperative experimental work of the Philippine Fisheries Commission and the author has been a *sine qua non* in respect to the successes obtained.

References

BERNATOWICZ, A. J. (1952). Seasonal aspects of the Bermuda algal flora. *Papers of Mich. Acad. of Sci., Arts and Letters*, 1950, **36**, 3–8.

BIEBL, RICHARD (1962). Temperaturresistenz tropischer Meeresalgen. *Botanica Marine* **4**, 3/4, 241–254.

BOERGESEN, F. (1911). The algal vegetation of the lagoons in the Danish West Indies. *Biologiske Arbeider Tilegnede Eug. Warming*. Pp 41–55.

CHANDLER, D. C. (1937). Fate of typical lake plankton in streams. *Ecological Monographs*, **7**.

COOPER, L. H. N. (1938). Redefinition of the anomaly of the nitratephosphate ratio. *Marine Biol. Association of the U. K., Jour.* **23**, 179f.

DOTY, M. S. (1946). Critical tide factors that are correlated with the vertical distribution of marine algae and other organisms along the Pacific Coast. *Ecology* **27**, 4, 315–328.

DOTY, M. S., and MORRISON, J. P. E. (1954). Interrelationships of the organisms on Raroia aside from man. *Atoll Research Bull.* **35**, 1–61, 9 figs.

EARLE, SYLVIA A. (1969). Phaeophyta of the eastern Gulf of Mexico. *Phycologia* **7**, 71–254.

FELDMANN, JEAN (1938). Recherches sur la vegetation marine de la Mediterranee. *Revue Algologique* **10**, 1–339.

GESSNER, FRITZ VON (1950). Die ökologische Bedeutung der Strömungsgeschwindigkeit fließender Gewässer und ihre Messung auf kleinstem Raum. *Sonderdruck aus dem Archiv für Hydrobiologie* XLIII: 159–165.

GHELARDI, R. J., and NORTH, W. J. (1958). A possible ecological effect of upwelling in a submarine canyon. *Nature* **181**, 207–208.

HAUG, ARNE, and JENSEN, ARNE (1954). Seasonal variations in the chemical composition of *Alaria esculente, Laminaria saccharinea, Laminaria hyperborea* and *Laminaria digitata* from *Northern Norway. Norwegian Institute of Seaweed Research*, Report no. **4**, 20 pp.

HAYASHIDA, F., and SAKURAI, T. (1969). Algal flora and communities at Mochimune, Suruga Bay. *Jap. J. Ecol.* **19**, 52–56.

HIATT, R. W. (1957). Factors influencing the distribution of corals on the reefs of Arno Atoll, Marshall Islands. *Proceedings of the 8th Pac. Sci. Cong.* **IIIA**, 929–970.

JITTS, H. R., MCALLISTER, C. D., STEPHENS, K. and STRICKLAND, J. D. H. (1964). The cell division rates of some marine phytoplankters as a function of light and temperature. *Jour. Fish. Res. Bd. Canada* **21**, 1, 139–157.

JONES, W. E., and DEMETROPOULOS, A. (1968). Exposure to wave action: measurements. of an important ecological parameter on rocky shores on Anglesey. *Jour. of Experimental Marine Biology and Ecology* **2**, 46–63.

KRISHNAMURTHY, V. (1967). Marine algal cultivation—necessity, principles and problems. *Proceedings of the Seminar on Sea, Salt and Plants*, 327–333.

LAWSON, G. W. (1957a). Some features of the intertidal ecology of Sierra Leone. *J. West African Science Ass.* **3**, 2, 166–174.

LAWSON, G. W. (1957b). Seasonal variation of intertidal zonation on the coast of Ghana in relation to tidal factors. *J. Ecol.* **45**, 831–860.

LAWSON, G. W. (1959). Application of analysis of variance to problems of intertidal ecology. (Abstract) *Proc. IX International Bot. Cong., Montreal*, Vol. **II**, 217.

LAWSON, G. W. (1966). The littoral ecology of West Africa. *Oceanogr. Mar. Biol. Ann. Rev.* **4**, 405–448.

LIST, ROBERT J. (1966). Smithsonian Meterorological Tables. *Smithsonian Miscellaneous Collections* 114, pp. xi + 527.

MCMICHAEL, DONALD F. (1966). Marine national parks in Australia, present developments and future needs. Special Symposium No. 1: Marine Parks, Sept. 4–8, 1966, *11th Pac. Sci. Cong.*, Tokyo, pp. 1–14.

NORTH, WHEELER J. (1964). Ecology of the rocky nearshore environment in Southern California and possible influences of discharged wastes. *International Conf. on Water Pollution Research, London, Sept. 1962.* Pp. 247–274. Pergamon Press, London.

OTT, JÖRG (1967). Vertikalverteilung von Nematoden in Beständen nordadriatischer Sargassaceen. *Helgolander wissenschaftliche Meeresuntersuchungen* 15, 1–4, 412–428.

RUTTNER, F. (1926). Bemerkungen über den Sauerstoffgehalt der Gewässer und dessen Respiratorischen Wert. *Naturwissenschaften* 14, 1237–1239.

RYTHER, J. H. and HULBURT, E. M. (1960). On winter mixing and vertical distribution of phytoplankton. *Limnol. and Oceanogr.* 5, 337–338.

RYTHER, J. H. and MENZEL, D. W. (1960). The seasonal and geographical range of primary production in the Western Sargasso Sea. *Deep-Sea Research* 6, 235–238.

SUBBARAMAIAH, K., and KRISHNAMURTHY, V. (1967). Laboratory culture of seaweeds. *Proceedings of the Seminar on Sea, Salt and Plants*, 321–326.

SVEDELIUS, NILS (1906a). Ecological and systematic studies of the Ceylon species of *Caulerpa. Ceylon Marine Biological Reports Part II.* I, 4, 81–144.

SVEDELIUS, NILS (1906b). Ueber die Algenvegetation eines ceylonischen Korallenriffes mit besonderer Ruecksicht auf ihre Periodizitaet. *Botaniska Studier tillaegnade F. R. Kjellman*, pp. 184–220.

WELCH, PAUL S. (1948). *Limnological Methods.* Xviii + 381 pp. The Blakiston Company, Philadelphia.

WHITFORD, L. A. (1960). The current effect and growth of fresh-water algae. *Transactions of the American Microscopical Society* 79, 302–309.

YENTSCH, C. S., and VACCARO, R. F. (1958). Phytoplankton nitrogen in the ocean. *Limnol. and Oceanogr.* 3, 443–448.

Coastal upwelling and sinking in the Caribbean Sea— especially about the existence of the under current

JIRO FUKUOKA

Universidad de Oriente
Instituto Oceanográfico
Cumaná, Venezuela

Abstract

In this work the mechanism of the upwelling, sinking and the counter under current in the Caribbean Sea may be explained by the action of the wind effect and the existence of the western boundary. The strongest wind (east-west component) occurs near 13° N in the Caribbean Sea. When the condition of western boundary and the wind distribution are utilized for the calculation of geostrophic transport in the Caribbean Sea, we can find the eastward geostrophic transport until 14° N from the continent, and this transport is proportional from the distance of the western boundary. On the other hand, the divergence of the Ekman transport is compensated by the convergence of the geostrophic transport under the supposition of stationary state. According to the distributions of convergence and divergence of the Ekman and geostrophic transports the north-south gradients of temperature and density between surface and sub-surface layers have different directions. Thus, the different currents may be esti-

123

mated in both layers. The existence of the eastward under current may be explained by the western boundary condition and the compensation between the Ekman transport and the geostrophic transport.

Resumen

Neste trabalho, o mecanismo da ressurgência, afundamento e a contra-corrente no Mar das Caraibas, pode ser explicado pela ação do efeito dos ventos e a existencia de um limite oeste. O componente mais forte (leste-oeste) ocorre próximo a 13° N no Mar das Caraibas. Quando as condições do limite oeste e a distribuição do vento são utilizadas para o cáculo do transporte geostrofico no Mar das Caraibas, pode verificar-se que êste ocorre até 14° N e é proporcional ao limite oeste da convergência.

De acordo com as distribuições de convergência e divergência dos transportes de Ekman e geostróficos, os gradientes norte-sul de temperatura e densidade, entre a superfície e sub-superfície, têm direções diferentes. Assim, as diferentes correntes podem ser estimadas em ambas as camadas.

INTRODUCTION

Oceanographic investigations have been conducted in the Caribbean Sea since the 1930's. One of the first studies, made by Parr (1937), used data collected by the research ship Atlantis I during 1933 and 1934 cruises. Of the recently published papers, the ones by Wüst (1963, 1964) are noteworthy because of the wide scope of analysis of the general oceanography.

This paper is concerned with the problems of upwelling and the current system in the Caribbean Sea. Upwelling in the Caribbean Sea has been studied by several oceanographers (Parr, 1937; Fukuoka, 1964, 1965; Ljøen and Herrera, 1965; Perlroth, 1968). In this paper, the author explains upwelling, sinking and under current by the existence of the western boundary in the Caribbean and the distribution of winds. It is also proposed that the under current may be the result of compensation between—the divergence of the Ekman transport and the geostrophic transport.

DISTRIBUTIONS OF GEOPOTENTIAL HEIGHTS

It is necessary to know the general features of ocean circulation in order to study the hydrographic condition in the Caribbean Sea. Unfortunately, the distribution of hydrographic stations is so wide that it is difficult to obtain a general pattern of currents. The geopotential heights reported by

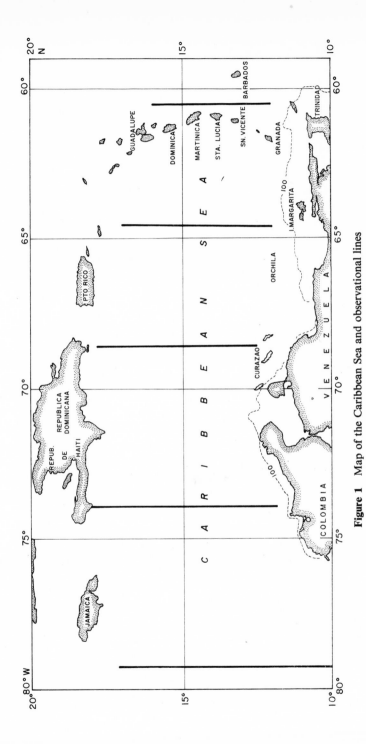

Figure 1 Map of the Caribbean Sea and observational lines

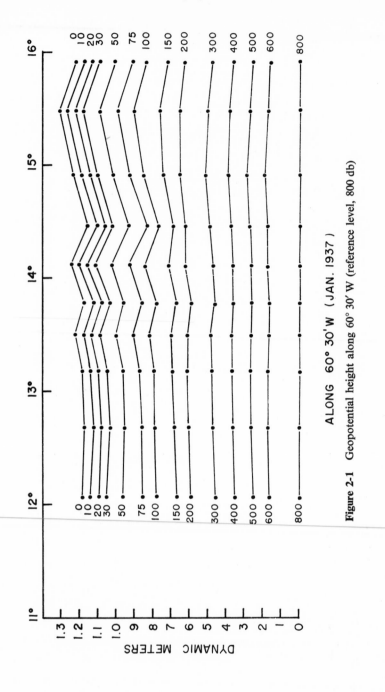

ALONG 60° 30'W (JAN. 1937)

Figure 2-1 Geopotential height along 60° 30' W (reference level, 800 db)

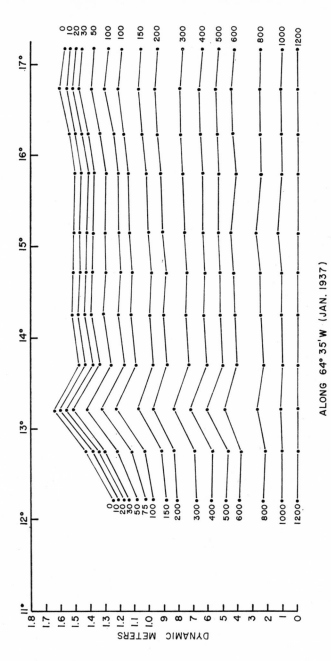

ALONG 64° 35'W (JAN. 1937)

Figure 2-2 Geopotential height along 64° 35' W (reference level, 1200 db)

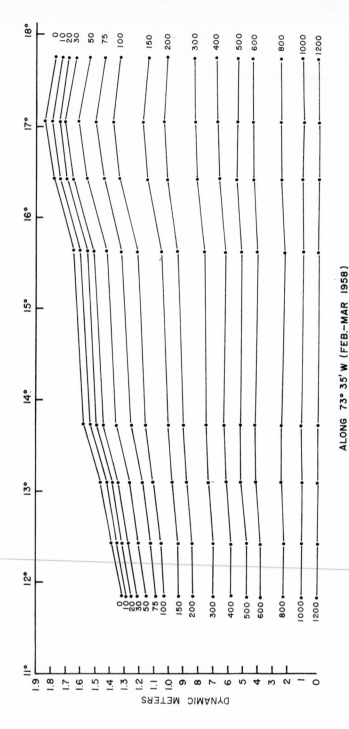

ALONG 73° 35' W (FEB.–MAR 1958)

Figure 2-3 Geopotential height along 73° 35′ W (reference level, 1200 db)

Woods Hole Oceanographic Institution cruises have been used to draw a general picture of current distribution (Fig. 1).

Since the currents in the Caribbean are dominently east-west, an analysis of currents can be made from north-south aligned sections of geopotential height (Fig. 2). The gradients of dynamic height for the sections in the east (60° 30′ W–64° 30′ W) indicate a complicated current structure. The section along 60° 30′ W shows westward currents at 14° N, 13° 20′ N and 15° N. Currents flowing toward the east are found at 13° 40′ N, 14° 20′ N and 15° 45′ N. Along the section 64° 30′ W the strong westward current is found 12° 45′ N but at 13° 20′ N there is a strong eastward current for the 1937 and 1958 data. The same section measured in 1933 shows the westward currents but no eastward currents except for a weak under current near 12° 30′ N.

In the area between 68° W and 79° W the ocean currents flowing west are observed near 12° 30′ N, 13° 30′ N, 14° N and 14° 40′ N, but they are not strong. There are no indications of the strong eastward currents in this area. Thus, when eastern and western parts in the Caribbean Sea are compared to each other in regard to current structure, it is apparent that there are more disturbances in the eastern part than in the western part.

DISTRIBUTIONS OF WATER TEMPERATURE AND DENSITY

Water temperatures are lower near the coast of the South American Continent. This low temperature may be due to coastal upwelling and the influence of the Sub-Antarctic intermediate water. Coastal upwelling in waters off eastern Venezuela has been reportes by Fukuoka (1964, 1966). Ljøen and Herrera (1965) reported upwelling near the Peninsula Paria and Perlroth (1968) reported upwelling near Isla Margarita and the Peninsula Guajira (near 72° W), but sometimes in the same place with low temperature near surface, relatively high temperature is found at some depths (150 m or 200 m). These distributions are shown in Figures 3 and 4. Areas in which low temperature appears near the surface and high temperature is seen at some depths (150 m or 200 m) in the same place are found along the eastern coast of Venezuela (Fig. 4).

The vertical distribution of water temperature and density show ascending isotherms and isopycnals in the surface layer toward the south (Fig. 5). The distribution of isotherms and isopycnals indicate the existence of different currents between the surface and deep layer (subsurface).

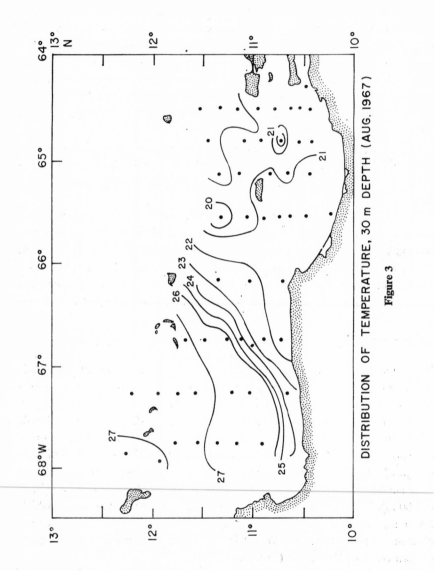

DISTRIBUTION OF TEMPERATURE, 30 m DEPTH (AUG. 1967)

Figure 3

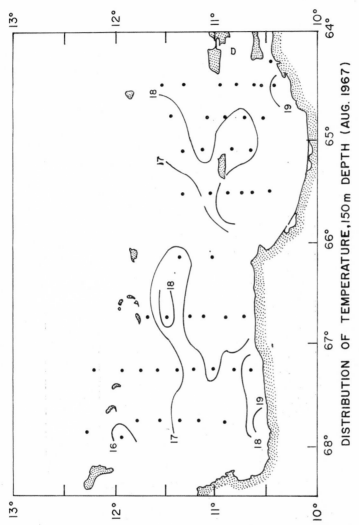

DISTRIBUTION OF TEMPERATURE, 150 m DEPTH (AUG. 1967)

Figure 3   Distributions of temperature, 30 m and 150 m depths

9*

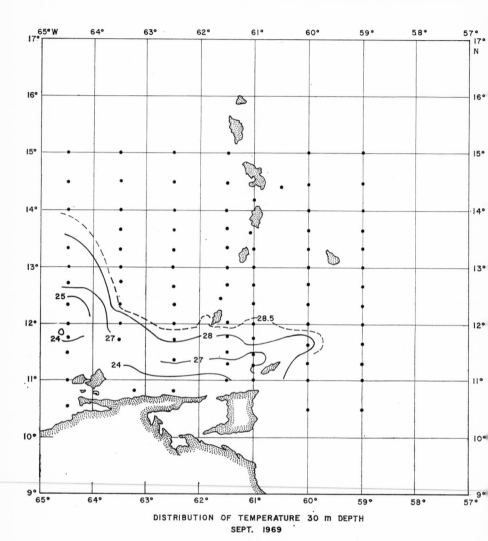

DISTRIBUTION OF TEMPERATURE 30 m DEPTH
SEPT. 1969

Figure 4-1 Distribution of temperature, 30 m depth

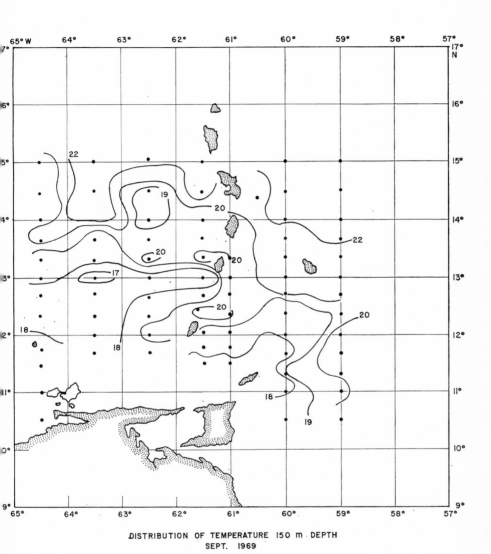

DISTRIBUTION OF TEMPERATURE 150 m DEPTH
SEPT. 1969

Figure 4-2 Distribution of temperature, 150 m depth

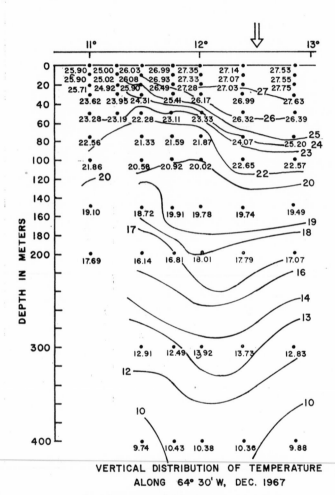

VERTICAL DISTRIBUTION OF TEMPERATURE
ALONG 64° 30' W, DEC. 1967

Figure 5 Vertical distributions of temperature and density (⇓ position of counter-undercurrent)

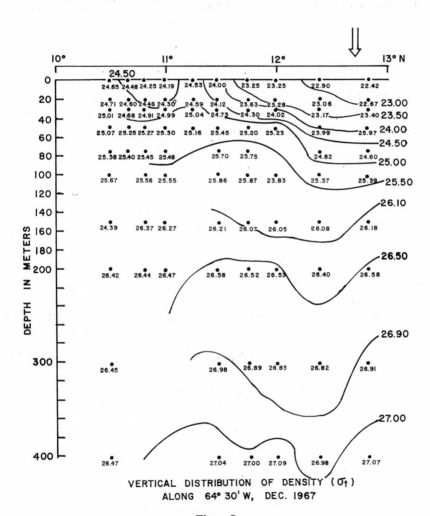

VERTICAL DISTRIBUTION OF DENSITY (σ_t)
ALONG 64° 30' W, DEC. 1967

Figure 5

ANALYSIS OF THE CURRENT IN THE SEA ADJACENT TO VENEZUELA

The vorticity equation of mass transport according to Munk (1950) is:

$$My = \frac{\text{curl}\,\tau}{\beta} \tag{1}$$

Where My is the north-south component of the mass transport, τ is the wind stress on the sea surface and β is latitudal variation of the Coriolis parameter. The X-axis is positive toward the east and the Y-axis is positive toward the north. Since τ_y is very small in the Caribbean Sea.

$$\text{curl}\,\tau = -\frac{\partial \tau_x}{\partial y}$$

The total mass transport may be divided into two parts; the Ekman transport and the geostrophic transport:

$$Mye + Myg = -\frac{1}{\beta}\frac{\partial \tau_x}{\partial y} \tag{2}$$

Where Mye is the north-south component of the Ekman transport and Myg is the geostrophic transport. The Ekman transport can be written as:

$$Mye = -\frac{\tau_x}{f} \tag{3}$$

Where f is the Coriolis parameter. Then from formulas (2) and (3):

$$Myg = -\frac{1}{\beta}\frac{\partial \tau_x}{\partial y} + \frac{\tau_x}{f} \tag{4}$$

The horizontal divergence of the geostrophic transport is written as:

$$\frac{\partial Myg}{\partial y} + \frac{\partial Mxg}{\partial x} = -\frac{\beta}{f} Myg \tag{5}$$

Differentiation of the formula (4) by y and the use of the formula (5) gives:

$$\frac{\partial Mxg}{\partial x} = -\frac{1}{\beta^2}\frac{\partial \beta}{\partial y}\frac{\partial \tau_x}{\partial y} + \frac{1}{\beta}\frac{\partial^2 \tau_x}{\partial y^2} \tag{6}$$

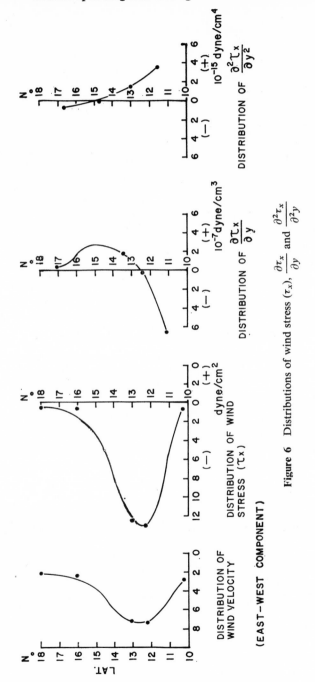

Figure 6 Distributions of wind stress (τ_x), $\dfrac{\partial \tau_x}{\partial y}$ and $\dfrac{\partial^2 \tau_x}{\partial^2 y}$

If R is the radius of the earth, φ is latitude and ω is angular velocity of the earth.

$$dy = Rd\varphi, \quad \beta = \frac{2\omega \cos \varphi}{R} \frac{\partial \beta}{\partial y} = -\frac{2\omega \sin \varphi}{R^2}$$

The distribution of wind and some values related to wind stress are shown in Figure 6, but these data are observed only in islands and continent. In Table 1 the distribution of wind stress show an average value of 5° squares of latitude and longitude (Hidaka, 1955). From these data the existence of the strongest wind stress may be estimated near 13° N in the Caribbean Sea. From Figure 6 the values of formula (6) along 13° N are given as:

$$-\frac{1}{\beta^2} \frac{\partial \beta}{\partial y} \frac{\partial \tau_x}{\partial y} \approxeq 0.2 \times 10^{-4} \quad \text{(c.g.s.)}$$

$$\frac{1}{\beta} \frac{\partial^2 \tau}{\partial y^2} \approxeq 0.75 \times 10^{-2} \quad \text{(c.g.s.)}$$

Table 1 Annual means of wind stress (East-west component, negative values correspond to westward, unit 10^{-2} dyne/cm^2, after Hidaka)

N°/W°	77.5	72.5	67.5	62.5	57.5	52.5
22.5	−45	− 56	−57	− 2	−49	− 60
17.5	−63	− 58	−70	−81	−79	− 82
12.5	−79	−124	−83	−65	−82	−107
7.5					−52	− 51

(‾ is maximum value on each column)

Thus, formula (6) is changed to:

$$\frac{\partial Mxg}{\partial x} \sim \frac{1}{\beta} \frac{\partial^2 \tau_x}{\partial y^2} \tag{7}$$

When formula (7) is integrated along x,

$$Mx_0 g - Mxg \sim \frac{1}{\beta} \frac{\overline{\partial^2 \tau_x}}{\partial y^2} (x_0 - x)$$

Where Mx_0g is zero on the western boundary (the Yucatan Peninsula) and $\dfrac{\partial^2 \tau_x}{\partial y^2}$ shows mean value along X.

Then:

$$Mxg \sim -\frac{1}{\beta}\overline{\frac{\partial^2 \tau_x}{\partial y^2}}(x_0 - x) \qquad (8)$$

Since there is only a western boundary, $(x_0 - x)$ is always negative, and $\dfrac{\partial^2 \tau_x}{\partial y^2}$ is positive at 14° N. Thus, Mxg becomes positive until 14° N and its value increases toward the east (Fig. 7). On the other hand, the eastward

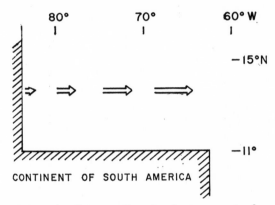

Figure 7 Schematic diagram of eastward component of geostrophic transport

under current may be explained the same way as the Cromwell current (Knauss, 1960, 1966) or an eastern boundary current system (Yoshida, 1959, Wooster and Gilmartin, 1961).

According to the distribution of wind, the divergence may be considered in the Ekman layer until 13° N and the convergence may be estimated north from 13° N. The convergence may be estimated in the geostrophic layer to 13° N and the divergence may appear north of 13° N. (Refer to distribution of wind on Figure 6.)

Thus, the tendencies of the iso-lines (isotherm or isopycnal) may be contrary between the surface layer and sub-surface layer due to the distributions of convergence and divergence (Fig. 8). From the distribution of the iso-line we can estimate the eastward current in the sub-surface.

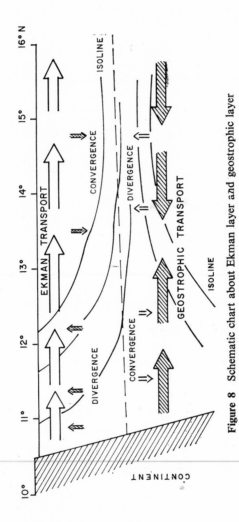

Figure 8 Schematic chart about Ekman layer and geostrophic layer

CONCLUSION

The current structure in the eastern part of the Caribbean Sea is complicated. There are areas of upwelling and sinking which depend upon the depth and of eastward under current.

These phenomena may be explained by the presence of the western boundary and the distribution of wind, and also by the balance of horizontal divergence between the Ekman layer and the geostrophic layer. Since the eastward under current is stronger in the east than in the west of the Caribbean Sea, the combination of eastward currents by the western boundary and by the compensation of both horizontal divergences may give a suitable explanation for these phenomena.

Acknowledgements

I am indebted to Dr. K. Yoshida who gave me valuable advice. Also I wish to express my gratitude to Dr. J. Morelock and Prof. L. Herrera who kindly polished English of manuscript.

References

FUKUOKA, J., BALLESTER, A. and CERVIGON, F. (1964). An analysis of hydrographical condition in the Caribbean Sea (III) especially about upwelling and sinking. *Studies on Oceanography*, Tokyo 145–149.

FUKUOKA, J. (1965). Coastal upwelling near Venezuela (1)—year to year change of upwelling. *Bol. Inst. Oceanogr.* **4**, 2, 223–233.

FUKUOKA, J. (1966). Coastal upwelling near Venezuela (II)—Certain periodicities of hydrographical conditions. *Bol. Inst. Oceanogr.* **5**, 1–2, 84–95.

HIDAKA, K. (1958). Computation of the wind stress over the oceans. *Rec. Oceanogr. Works*, Japan. **4**, 2, 77–123.

KNAUSS, J. A. (1960). Measurements of the Cromwell current. *Deep-Sea Res.* **6**, 2, 265–286.

KNAUSS, J. A. (1966). Further measurements and observations on the Cromwell current. *Jour. Mar. Res.* **24**, 2, 205–240.

LJØEN, R. and HERRERA, L. E. (1965). Some oceanographic conditions of the coastal waters of eastern Venezuela. *Bol. Inst. Oceanogr.* **4**, 1, 7–50.

MUNK, W. H. (1950). On the wind-driven ocean circulation. *Jour. Meter.* **7**, 79–93.

PARR, A. E. (1937). Contribution to the hydrography of Caribbean and Cayman Seas, based upon the observations made by the research ship Atlantis 1933–1934. *Bull. Bingham Oceanogr. Coll.* **5**, 4, 1–10.

PERLROTH, I. (1968). Distribution of mass in the near surface waters of the Caribbean. *Nat. Oceanogr. Data Center Progress Rep.* p. 72. Nov. 1–15.

WOOSTER, W. S. and GILMARTIN, M. (1961). The Perú-Chile undercurrent. *Jour. Mar. Res.* **19**, 3, 97–122.

Wüst, G. (1963). On the stratification and the circulation in the cold watersphere of the Antillean-Caribbean basins. *Deep-Sea Res.* **10**, 3, 165–187.

Wüst, G. (1964). *Stratification and circulation in the Antillean Caribbean basins.* Part. 1. Columbia Univ. Press.

Yoshida, K. (1955). Coastal upwelling off the California coast. *Rec. Oceanogr. Works., Japan.* **2**, 2, 1–13.

Yoshida, K. (1959). A theory of the Cromwell current (the Equatorial undercurrent) and of the Equatorial upwelling. An interpretation in a similarity to a coastal circulation. *Jour. Oceanogr. Soc. Japan.* **15**, 4, 159–170.

River–ocean interactions

EDWARD D. GOLDBERG

Scripps Institution of Oceanography
University of California at San Diego, La Jolla, California

Abstract

The introduction of the dissolved and suspended species in river waters to the marine environment has measurable impacts upon the composition of the coastal waters and upon the sediments for distances of hundreds of kilometers. Two general problems have been the focus of attention in the zones of river-ocean interactions: (1) the effect of river influents upon biological activity; and (2) the types of inorganic chemical reactions that take place between the dissolved and the suspended loads of the rivers and of the oceans.

Resumen

A introdução, nos rios, de espécies dissolvidas e em suspensão, causa, no ambiente marinho, consideráveis impactos na composição das águas costeiras e nos sedimentos, em centenays de quilômetros. Dois problemas gerais são o principal foco de atenção em zonas de interação rios/oceano:
1. A influência dos rio; nos processos biológicos.
2. Os tipos de tenções químicas inorgânicas, que tomam parte entre as cargas dissolvidas e em suspensão nos rios e oceanos.

INTRODUCTION

The transfer of materials from the continents to the oceans takes place primarily through rivers. Additional transport occurs in the wind systems and in the glaciers. Also, man has enhanced the redistributions of matter over the surface of the earth by his sewage outfalls and his sea going craft.

Two general types of problems have attracted the attention of environmental scientists in the province of river-ocean interactions:

A. What is the effect of the effluent of a river upon primary plant productivity, as well as upon other types of biological activity, in the marine domain?

B. What inorganic reaction takes place between the dissolved and suspended loads of the rivers and of the oceans?

Clearly, each river provides a unique environment for the study of such problems and many such studies have been made. Yet, the systematics of the biological and of the inorganic chemistries are not understood well. The strategy of this presentation will be to examine some of the more recent results, not exhaustively, but illustratively to point out where may be the more rewarding entries for future investigations.

Such studies are also of significance in predicting the fates of the man-produced pollutants that the rivers now carry. Many of the problems of waste management are dependent for solution upon information gained from the investigations involving phenomena in river-ocean systems.

SOME PRELIMINARY DATA

Yearly the rivers of the world bring into the oceans 3.9 billion tons of dissolved substances and 18 billion tons of suspended solids* (Holeman, 1968). Each posseses a unique composition that can vary widely as a function of time. The dissolved species can differ from river to river by two orders of magnitude or more. In spite of such variations, it is most rewarding to have a mean composition of river waters of the world, such as that prepared by Livingston (1963), for an initial entry to a study of river-ocean interactions.

Table 1 presents the principal species of seawater and of rivers. Although most of the substances are markedly accumulated in the oceanic domain, three elements normally considered as necessary for plant growth there, silicon, nitrogen and iron, have like or greater concentrations in the rivers.

---

* Recent estimates of this amount have varied between 10 and 60 billion tons. This value is presented as a reasonable estimate based upon a consideration of a somewhat more extensive set of data than was used by previous investigators.

Table 1 Mean composition of river water and that of sea water for a chlorinity of 19‰
Concentrations are given in milligrams/liter

Species	Ocean water content	River water content (Livingston, 1963)
Na^+	10,800	6.3
K^+	390	2.3
Ca^{++}	400	15
Mg^{++}	1,300	4.1
Fe (all forms)	0.01	0.67
Cl^-	19,000	7.8
SO_4^{--}	2,700	11.2
NO_3^-	3	1.0
SiO_2	6	13.1
HCO_3^-	130	58.4

Phosphorus, also considered an essential element for the growth of plants, represents a similar situation with levels of 400 parts per billion in streams and 88 parts per billion in seawater (Turekian, 1969). Sometimes, some nutrients are enriched, and others not, as appears to be the case with the Columbia River compared to the adjacent waters during the winters (Stefánsson and Richards, 1964):

Species (Micromoles/liter)	Columbia River Water	Surface Pacific Ocean Water
Silicate	208	0.4
Nitrate	25	5.5
Phosphate	1	1

On the other hand, all such species may be lower in the river waters as compared to the oceans. Williams (1968) came across such a situation in the Amazon:

Species (micrograms/liter)	Amazon River Water	Surface Atlantic Ocean Water
Nitrate	55–135	280
Phosphate	0.3–21.7	62
Silicon	12–1743	3000

The conclusion that appears reasonable from such data is that the mixing of river waters with seawater could either fertilize or hinder plant growth with respect to the availability of these nutrients.

Most of the minor elements have higher concentrations in the rivers than in the oceans. Table 2, with data derived from Turekian (1969) classifies these elements on the basis of their concentrations in seawater and rivers: the first group has those elements which have the concentration ratio, $C_{\text{seawater}}/C_{\text{river water}}$, ten or higher, while the second column has those elements with a concentration ratio of less than 10. This second group encompasses the more reactive elements, those that readily engage in the inorganic or biochemical processes in the ocean. The first group contains the relatively unreactive elements, whose aqueous species prefer the dissolved state, such as the alkali and alkaline earths and the halogens.

Table 2 Trace elementes in rivers and the oceans. Data from
Turekian (1969)

Concentrations in parts per billion (micrograms/liter)

Element	River	Ocean	Element	River	Ocean
Li	3	170	Al	400	1
B	10	4450	Ti	3	1
F	100	1300	V	0.9	1.9
Ni	0.3	6.6	Cr	1	0.2
Br	20	67,300	Mn	7	0.4
Rb	1	120	Co	0.2	0.39
Sr	50	8100	Cu	7	0.9
Mo	1	10	Zn	20	5
U	0.04	3.3	Ga	0.09	0.03
			As	2	2.6
			Se	0.2	0.09
			Y	0.7	0.013
			Ag	0.3	0.28
			I	7	64
			Cs	0.2	0.30
			Ba	10	21
			La	0.2	0.0034
			W	0.3	<0.001
			Au	0.002	0.011
			Hg	0.07	0.15
			Pb	3	0.03
			Th	0.1	0.0004

The dissolved and particulate organic phases of the Amazon River (six different localities) were analyzed by Williams (1968) and the data are collated in Table 3. With the exception of the dissolved organic phosphorus the river contains larger amounts of organic substances than do surface seawaters. The high organic carbon values came from the Rio Branco–Rio Negro drainage area where the brownish yellow waters contain large amounts of humic acids.

Table 3 The dissolved and particulate species of the Amazon River (Williams, 1968)

	Dissolved Constituents (µg/l)		Particulate Constituents (µg/l)	
	Amazon River	Surface Seawater	Amazon River	Surface Seawater
Organic carbon	1640–6300	1000	740–2060	100
Organic nitrogen	130–212	100	87–173	10
Organic phosphorus	0.6–5.3	10	4–87	1

The supersaturation of river waters with respect to atmospheric carbon dioxide is exemplified by the Columbia River where an excess of the gas to levels of 200–870 parts per million is reported (Park *et al.*, 1969).

THE GEOGRAPHICAL DEFINITION OF THE RIVER–OCEAN INTER-ACTIONS

The river water injections into the oceans can be detected through decreases in the salinity or in the density of the seawater or through the dilution of a dissolved species whose concentration is markedly different in rivers, such as monomeric silicic acid. Recently, radioactive tracers derived from nuclear reactors have appeared especially attractive to determine the rates of transport and mixing of fresh and sea waters. Chromium-51, with a radioactive half-life of 28 days, has defined the Columbia River effluent (Fig. 1) to distances of 350 kilometers to sea (Osterberg *et al.*, 1965; Gross *et al.*, 1965). This nuclide appeared to be a preferable tracer to others such as Zinc-65 inasmuch as it remains in solution and is not appreciably concentrated by the biota. The introduction of the time parameter to river–ocean interactions was made by Osterberg *et al.* (*op. cit.*) who found that the maximum rate of plume movement was 11.4 km per day, based upon the radio-

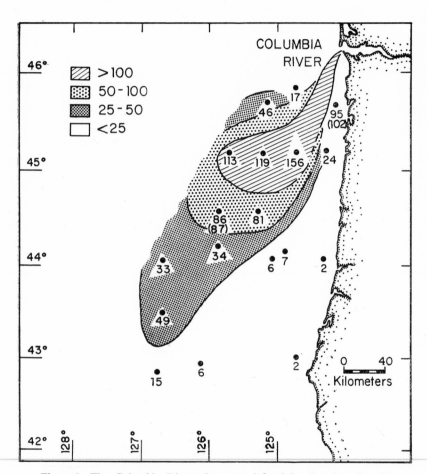

Figure 1   The Columbia River plume as defined by the chromium-51 activity of surface sea water (counts per minute per 100 liters of water). Numbers in parentheses indicate duplicate samples. From Osterberg *et al.* (1965)

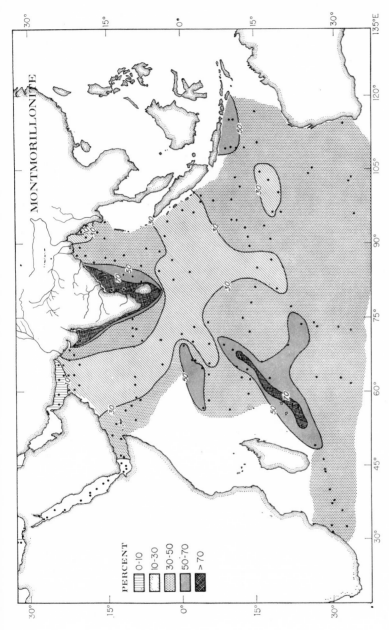

Figure 2 Montmorillonite concentrations in the under 2 micron size class of surface sediments from the Indian Ocean. Samples analyzed are indicated by dots. Unpublished data of Edward D. Goldberg

active decay of Cr-51. This value is a minimum speed inasmuch as any losses of Cr-51 by biological uptake or precipitation would decrease the apparent rate of flow. Physical techniques have measured the flow rate and give essentially the same result as the radioactive ones.

Mineral assemblages have been used to define the influence of rivers upon the coastal deposits. Off the coast of South America, the Magdalena, Orinoco and the Amazon Rivers produce tongues of sediments high in the clay mineral illite (Griffin *et al.*, 1968). The Indian peninsula is outlined as far as one thousand kilometers off the coast by the montmorillonite contents of the sediments carried to the oceans primarily by the rivers. This montmorillonite arises from the weathering of the Deccan Traps, a volcanic formation composed of an augite basalt (Fig. 2).

The Hanford reactors on the Columbia River supplied time-clocks for measuring the rate of movement of sediment along and away from the coast (Gross and Nelson, 1966). Zn-65 and Co-60 become incorporated in the sedimentary particles and are not readily removed by ion exchange processes. In a simple model to measure the rate of sediment movement it was assumed that the change in radioactivity with increasing distance from the river mouth is solely due to radioactive decay and not due to mixing with previously deposited sediments. A second model is based on the variations in the activity ratio of these two isotopes. The results indicates a movement of 12 to 30 kilometers per year along the shelf and 2.5 to 10 kilometers per year westward away from the coast.

Inorganic reactions

On the basis of both field and laboratory investigations, Bien *et al.* (1958) argued that soluble silica (most probably monomeric silicic acid $Si(OH)_4$) is removed from river waters entering the oceans by inorganic reactions involving the suspended matter. Surface seawaters in the Gulf of Mexico area where the Mississippi River enters show no marked enrichment of dissolved silicon. The river water contain 4 to 7.5 parts per million of silica while the average concentration in Gulf of Mexico surface water is 0.11 parts per million. Both biological removal and inorganic precipitation had been proposed to account for any depletion of dissolved silica.

If only dilution of the introduced dissolved silica takes place, then its concentration in the Gulf waters would be an inverse, linear function of the chlorinity. Measurements made in the area of the eastern Mississippi

River Delta at Pass a Loutre, where the chlorinities vary between 0.1 and $21^o/_{oo}$ indicated silica levels significantly different from those expected under a concept of dilution (Fig. 3). The deviations cannot be explained on the basis of experimental error. The work was carried out in the months of June, July and October and the phenomenon was evident during these three periods.

Laboratory experiments established that the suspended material from the river was a more effective scavenger of dissolved silica than either bentonite or aluminium oxide. Also, the necessity of electrolytes as well as the solid phases for the removal of silica was established. As the chlorosity is increased from 0 to 1 grams of chloride per liter, the removal of silica increased from 0 to about 8 % and further increases of chloride up to 4.3 grams/liter did not increase the sorption of silica on the particles. Increasing removal of silica occurs with increasing amounts of solids, indicating adsorption or coprecipitation of the monomeric silicic acid is taking place.

Biological incorporation of the dissolved silica in phytoplankton may be responsible for a part of the depletion in the Gulf of Mexico. Between 0.4 and 2.8 micromoles of dissolved silica are taken up per liter per day by the photosynthesizing biomass. This works out to be 2 percent of the total soluble silica in river water or about 20 percent of the silica depletion, assuming the biological activity takes place in one day. Such a situation could occur during periods of high discharge.

Other investigators, working in different localities have not been able to confirm a silica removal such as Bien *et al.* (1958) describe. In the low chlorinity waters of the Baltic (less than $5^o/_{oo}$) the silicate levels appear to be a direct consequence of the mixing of river water and ocean water without any biological or inorganic removal (Voipio, 1961).

On a much larger geographic scale, Schink (1967) finds no evidence for silica removal as rivers enter the Mediterranean Sea. A silica flux through the Straits of Gibralter is calculated to be 3.6×10^3 moles per second with an estimated influx from rivers and the Black Sea as 3.9×10^3 moles per second. This agreement strongly suggests that no substantial fraction of the dissolved silica is lost to the Mediterranean system.

During the winter season the waters of the Columbia River plume, according to Stefánsson and Richards (1964), are two component mixtures of Columbia River water and of surface sea water. There is no evidence for any inorganic precipitation of silica. These workers assumed that the

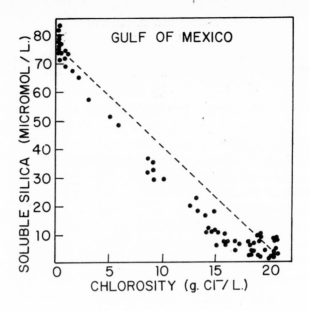

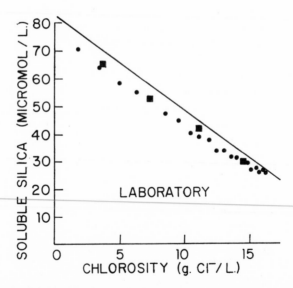

Figure 3 Soluble silica in the Gulf of Mexico as a function of chlorosity of water samples (upper). Soluble silica concentrations as a function of chlorosity as unfiltered river-water is mixed with Pacific Ocean water in the laboratory (lower). From Bien *et al.* (1958)

suspended load borne by the Columbia River was too small to effect any precipitation of silica when it entered the marine environment. However, during spring, summer and fall there were departures from predicted concentrations based upon simple mixing of river and ocean water, most probably as a result of subsurface additions of nutrient-rich waters and of biolgical activity in removing the silica.

The distribution of dissolved silicate is governed principally by the mixing processes in the Southampton Water, an estuary studied by Banoub and Burton (1968). During the winter months when the study was carried out, the biological influences on the distribution were assumed to be insignificant. The salinity range of the samples, 22 to 34°/₀₀, does not preclude a removal of silicon from solution in the earlier stages of the mixing of the river with the sea waters. On the average there was 2.5 times as much silicon present in particulate forms as was present in solution in these waters.

Biological productivity

The introduction of river water to the marine environment may exert an influence upon the primary plant productivity through effects upon any of the following parameters (adapted from Ryther *et al.*, 1967):

A. Nutrient species. If the plant nutrients are in higher concentrations in the river waters than in the sea, stimultation of plant growth may occur. If such substances are present in lower levels in river as compared to ocean water, the mixed waters will yield smaller plant populations. Further, river borne suspended loads may remove nutrient species from the waters to which they are introduced and hence limit plant activity.

B. Light intensity. Particulate species or colored dissolved substances can diminish the light intensity and hence the depth of the photic zone.

C. Stability of the photic zone. The formation of a low density surface layer increases the stability of the water column. As a result, phytoplankton populations can attain higher levels by a reduction of the probability that plants are carried below the critical depth for photosynthetic activity by turbulence.

The Amazon River waters have been traced by Ryther *et al.* (1967) into the Atlantic through their high silicate levels and through their diluting effect upon sea waters. The concentrations of the nutrient species of phosphorus and nitrogen were extremely low, perhaps of the level, or even less than those of typical tropical surface waters. In this low salinity region, the phytoplanktonic cell counts were smaller than those in adjacent waters suggesting that the direct impact of the river water is to lower the fertility of the marine region.

Shoreward of this low salinity lens, the phosphate levels and the total phytoplankton populations increase markedly. This biological activity was attributed to the geostrophic uptilting of the water in the region such that nutrient rich subsurface water was brought into the photic zone, a phenomenon perhaps related to the weight of the accumulated river water offshore.

Calef and Grice (1967) characterized these low salinity lenses by their high populations of two species of coastal zooplankton, the cladoceran *Evadne tergestina* and the decapod *Lucifer faxoni*. The sources of these organisms could not be defined. They may have been present in the inshore waters and persisted after the low salinity lenses were formed or they may have been produced following the entrainment of stray individuals in such waters.

This Amazon effluent has been observed to transport a group of neritic diatoms far off the coasts of Brazil, French Guiana, and Dutch Guiana during the fall season (Hulburt and Corwin, 1969). The water mass apparently crosses the northwestward Guiana Current and becomes entrained in a southeastward countercurrent. During the spring there was a much less abundant flora of these coastal organisms either because their supply was limited or the drift may have been too long and not appropriate for survival. The oceanic flora, predominantly coccolithophores, do very poorly in the Amazon influenced waters.

The greater surface stability in waters of the Columbia River plume as compared to the adjacent ocean waters appears to give rise to higher concentrations of diatoms and microflagellate (Hobson, 1966) in an investigation conducted during the winter. Although the primary productivity of both domains of water were quite similar, the diatom populations were substantially higher in the plume areas. It is of interest to note that the turbidity of the freshwaters may have depressed the light penetration enough that the total standing crop was smaller than would have been the case with more transparent water. There was no evidence for limitation of plant growth through inadequate supplies of nutrients. In the plume waters the depth of the mixed layer was negligible.

DISCUSSION

The inorganic chemistries and the biological activities that result from the mixing or river and ocean waters may increase our understanding of some of the fundamental problems of marine science. For example, the factor or factors that govern the concentration of dissolved silicon in the oceans are not as yet defined. Some investigators argue that the undersaturation of soluble silicate species with respect to such solid phases as quartz and opal results from the deposition of siliceous test by the photosynthesizing plants of the sea. Other argue that the silicon concentrations are determined by sorption reactions on detrital minerals introduced from the continents.

A resolution of this difficulty through investigations in coastal zones of high biological productivity and of high rates of sedimentation, such as the general areas where the rivers flow into the oceans, seemed reasonable. Yet the initial results of such work are conflicting, with the majority of workers not finding silica removal by sorption upon inorganic particles. Perhaps, different river systems and different seasons do provide the variety of results that are now found in the literature. Generalizations about the silica budget in the oceans may have to await studies on a greater number of rivers.

On the other hand, it may be rewarding to look at some other elements, especially those that have higher concentrations in the rivers, in the river–ocean mixing zones with regard to removal processes. Somayajulu and Goldberg (1966), for example, found two orders of magnitude more thorium in coastal waters than in the open ocean. The cerium contents of several eastern U. S. coastal water samples were between 0.1 and 0.6 micrograms/l while those in the open ocean near Bermuda were between 0.004 and 0.014 (Carpenter and Grant, 1967). Or again, Topping (1969) finds cobolt, iron and zinc in higher concentration in inshore waters than in those from the open ocean. What are the processes responsible for such distributions?

Similarly, the biological activities in river–ocean water admixtures do not seem to relate to chemical components of the systems. Variations in primary productivity between plume and adjacent waters have not as yet been observed, but might exist. Profitable investigations might be made in river systems where there are marked differences between the fresh and salt waters with respect to nutrients levels. Also, the influence of some of the minor elements, whose concentrations are greater in the rivers, upon primary productivity might be sought.

References

BANOUB, M. W. and BURTON, J. D. (1968). The winter distribution of silicate in Southampton Water. *J. Cons. perm. int. Explor. Mer.* **32**, 201–8.

BIEN, G. S., CONTOIS, D. E. and THOMAS, W. H. (1958). The removal of soluble silica from fresh water entering the sea. *Geochim. Cosmochim. Acta* **14**, 35–54.

CALEF, G. W. and GRICE, G. D. (1967). Influence of the Amazon River outflow on the ecology of the western tropical Atlantic. II. Zooplankton abundance, copepod distribution, with remarks on the fauna of the low-salinity area. *J. Marine Res.* **25**, 84–94.

CARPENTER, J. H. and GRANT, V. E. (1967). Concentration and state of cerium in coastal waters. *J. Marine Res.* **25**, 228–38.

GRIFFIN, J. J., WINDOM, H. and GOLDBERG, E. D. (1968). The distribution of clay minerals in the world ocean. *Deep-Sea Research* **15**, 433–59.

GROSS, M. G. and NELSON, J. L. (1966). Sediment movement on the continental shelf near Washington and Oregon. *Science* **154**, 879–81.

GROSS, M. G., BARNES, C. A. and RIEL, G. K. (1965). Radioactivity of the Columbia River Effluent. *Science* **149**, 1088–90.

HOBSON, L. A. (1966). Some influences of the Columbia River effluent on marine phytoplankton during January 1961. *Limnology Oceanogr.* **11**, 223–34.

HOLEMAN, J. N. (1968). The sediment yield of the major rivers of the world. *Water Resources Research* **4**, 737–47.

HULBURT, E. M. and CORWIN, N. (1969). Influence of the Amazon River outflow on the ecology of the western tropical Atlantic. III. The plankton flora between the Amazon River and the Windward Islands. *J. Marine Res.* **27**, 55–72.

LIVINGSTON, D. A. (1963). The chemical composition of rivers and lakes. *U. S. Geological Survey Prof. Paper 440G*, 64 pp.

OSTERBERG, C., CUTSHALL, N. and J. CRONIN (1965). Chromium-51 as a radioactive tracer of Columbia River water at sea. *Science* **150**, 1584–7.

PARK, K., GORDON, L. I., HAGER, S. W. and CISSELL, M. (1969). Carbon dioxide partial pressure in the Columbia River. *Science* **166**, 867–8.

RYTHER, J. H., MENZEL, D. W. and CORWIN, N. (1967). Influence of the Amazon River outflow on the ecology of the western tropical Atlantic. I. Hydrography and nutrient chemistry. *J. Marine Res.* **25**, 69–83.

SCHINK, D. R. (1967). Budget for dissolved silica in the Mediterranean Sea. *Geochim. Cosmochim. Acta* **31**, 978–99.

SOMAYAJULU, B. L. K. and GOLDBERG, E. D. (1966). Thorium and uranium isotopes in seawater and sediments. *Earth Planet. Sci. Letters* **1**, 102–6.

STEFÁNSSON, U. and RICHARDS, F. A. (1963). Processes contributing to the nutrient distributions off the Columbia River and Strait of Juan de Fuca. *Limnology Oceanogr.* **8**, 394–410.

TOPPING, G. (1969). Concentrations of Mn, Co, Cu, Fe, and Zn in the northern Indian Ocean. *J. Marine Res.* **27**, 318–26.

TUREKIAN, K. K. (1969). The oceans, streams and atmosphere. In *Handbook of Geochemistry* pp. 297–323 (ed. Wedepohl, K.) Berlin: Springer-Verlag.

VOIPIO, A. (1961). The silicate in the Baltic Sea. *Ann. Acad. Scient. Fennicae* **106A**, 1–15.

WILLIAMS, P. M. (1968). Organic and inorganic constituents of the Amazon River. *Nature* **218**, 937–8.

Primary productivity and phytoplankton in the coastal Peruvian waters

OSCAR GUILLÉN,
BLANCA ROJAS DE MENDIOLA,
and
RAQUEL IZAGUIRRE DE RONDÁN

Instituto Del Mar
Lima, Peru

Resumen

Con el objeto de conocer cuales son las áreas productivas y no productivas en las aguas costeras del Perú, se tomaron muestras a los niveles de 0–10–20 y 40 m para medir la capacidad de producción de las aguas, siguiendo la técnica descrita por Berge (1958), y expuestas a una iluminación de 10,000 luxes.

Se hicieron experimentos durante las estaciones de verano, otono, invierno y primavera del ano 1964, cuyos resultados se discuten y se relacionan con los datos del medio ambiente fisico-quimico, asi como con la producción primaria, obtenidos de acuerdo al método de Steemann Nielsen (1952). Igualmente se muestra la variación estacional del fitoplancton al nivel de 10 m y su composición.

Las áreas de mayor capacidad de producción estuvieron asociadas con los mayores valores de fitoplancton, y fueron hallados al sur de la latitud 6° S, a lo largo y cerca de la costa, favorecidos por el abastecimiento de nutrientes de los procesos de afloramiento.

Se halló qua la capacidad de producción a 10 m puede ser una buena medida de la producción total y producción primaria a 0 m.

INTRODUCTION

The coastal waters off Peru are among the most fertile in the world. A knowledge of its phytoplankton standing crop is, therefore, of great importance. During oceanographic cruises in 1964 phytoplankton studies were included not only because of the productivity indices obtained but also because phytoplankton is one of the favorite foods of anchoveta.

The results of the four 1964 seasonal cruises between 4 and 12° S latitudes are included here to present additional information about distribution and seasonal variations of the plankton and of the productive capacity of Peruvian coastal waters.

Barreda (1957), Rojas de Mendiola (1958, 1963 a, b, c) found seasonal variations both in the quantitative and the qualitative composition of the plankton. Recently Strickland *et al.* (1969) have made simultaneous studies of phytoplankton composition, rate of primary production, and nutrients.

MATERIAL AND METHODS

The 1964 seasonal "Unanue" cruises, which obtained data on phytoplankton and production capacity at 10 m, were as follows:

Summer cruise   6402; Feb. 25 to March 18
Autumn cruise   6405; May 30 to June 11
Winter cruise   6408; Aug. 17 to Sept. 11
Spring cruise   6411; Nov. 11 to Dec. 1

Each cruise included, in addition to the regular water sample drawn for salinity, oxygen and phosphate analysis, samples at 10 m for plankton analysis and C^{14} analysis. Samples were drawn at 10 m using Van Dorn type bottles (Fig. 1 gives the location of the C^{14} samples).

Productivity was measured by the radio carbon C^{14} method after Steemann Nielsen (1952) with the technique of Berge (1958). To each sample 4 μc of radioactivity were added prior to being illuminated at 10,000 luxes for 4 hours. The resulting productivity is expressed as mg C $\times$ 10^{-7}/L/Lux-h. Plankton samples for counts were preserved in neutral formalin (4 ml of a 30% solution per 100 ml of sample). At the laboratory the samples were processed following the sedimentation method of Vermohl (1936); sedimentation lasted 48 hours in cylinders of 10 and 25 ml. Samples were examined with zonar oculars of 8× and objectives of 10×. All the organisms deposited in one half of the cylinder base were counted in alternate zones.

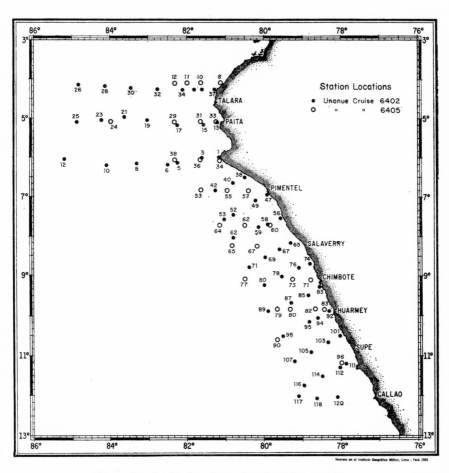

Figure 1a   Station locations of productivity capacity

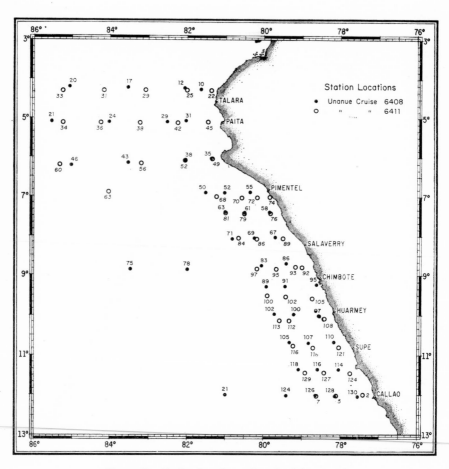

Figure 1b　Station locations of productivity capacity

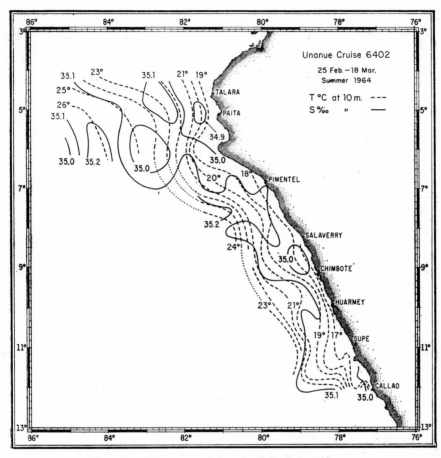

Figure 2a Temperature (--) and salinity (—) at 10 m

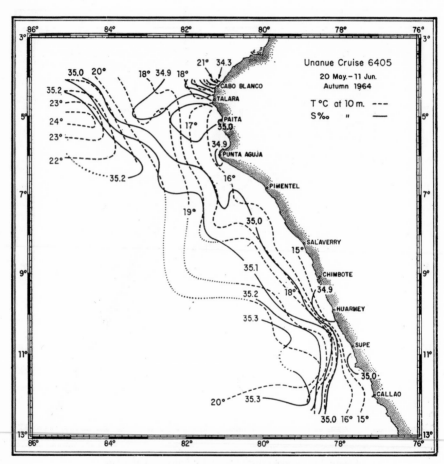

Figure 2b Temperature (--) and salinity (—) at 10 m

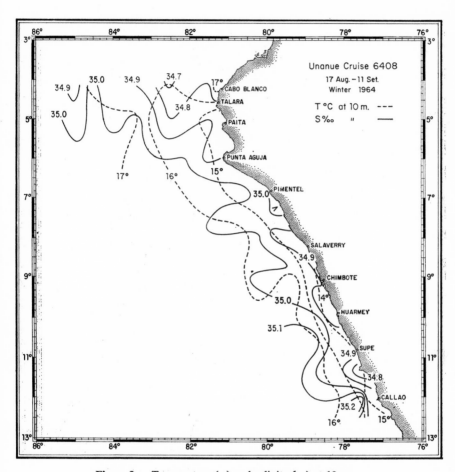

Figure 3a Temperature (--) and salinity (—) at 10 m

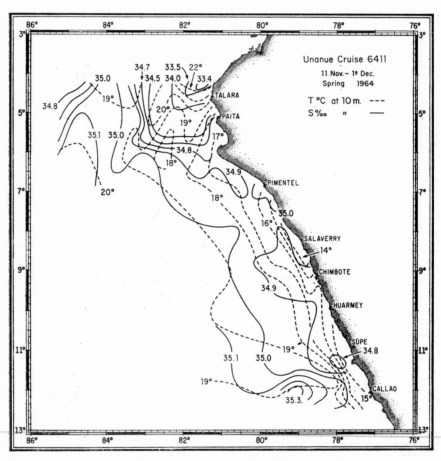

Figure 3b Temperature (--) and salinity (—) at 10 m

HYDROGRAPHIC CONDITIONS

The physical–chemical conditions of the coastal waters in 1964 (Figs.2 and 3) were presented by Zuta y Guillèn (1964) from which the summary pertinent to the 10 m depth is given here.

Temperature and salinity variations (10 m) were as follows:

	Summer	Autumn	Winter	Spring
T °C	26–17	24–15	17–14	22–14
S $^o/_{oo}$	35.2–34.9	35.3–34.3	35.2–34.7	35.3–33.4

Notable changes occur from summer to winter, especially in regard to temperature due to seasonal cooling and variations in the horizontal and vertical flow. The surface subtropical waters, identified by salinities greater than $35.1^o/_{oo}$, were present north of 13° S during all the year but were more pronounced during the summer and autumn. The major seasonal changes in salinity occurred north of Punta Falsa (6° S) where waters from the equatorial zone ($34.8-33.8^o/_{oo}$) are normally observed (see autumn and winter charts). Pronounced changes occurred during the spring due to the advance of surface tropical waters with salinities as low as $33.4^o/_{oo}$ which were a prelude to the "El Niño" phenomenon that took place during the summer of 1965.

RESULTS AND DISCUSSION

Distribution of the production capacity at 10 m

During the summer (Fig. 4) high productivity values were observed along and close to the coast South of 6° S Lat. with an average of 11.42 mg C × 10^{-7}/L/Lux-h. The main areas of productivity were found off Pimentel and Salaverry, reaching maximum values of 66.3 and 37.7 mg C × 10^{-7}/L/Lux-h, respectively. The upwelling area off Huarmey-Supe showed low productivity values near the coast because of water recently overturned that contained low phytoplankton cells. Far from the coast, however, an area of great productivity was encountered because an anticyclonic eddy had developed which favored the rapid development of phytoplankton.

Low productivity values were encountered in waters far away from the coast, west of the 82° W Long. in the northern region.

In the autumn (Fig. 4) productivity values were lower than in the summer because stability of the waters decreased and the light was not adequate.

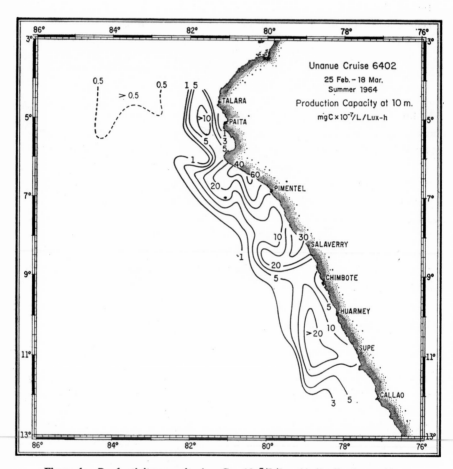

Figure 4a Productivity capacity (mg C $\times$ 10^{-7}/L/Lux-h) distribution at 10 m

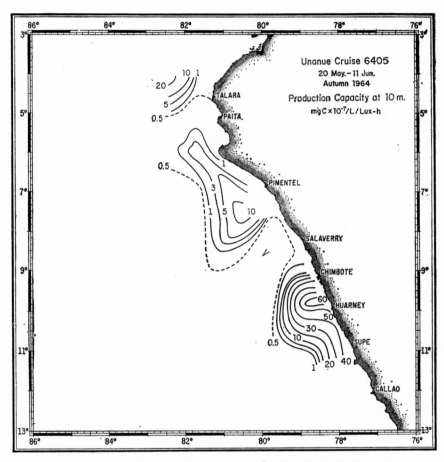

Figure 4 b   Productivity capacity (mg C $\times$ 10^{-7}/L/Lux-h) distribution at 10 m

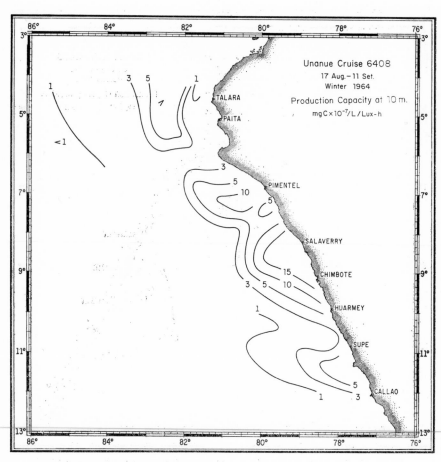

Figure 5a   Productivity capacity (mg C $\times$ 10^{-7}/L/Lux-h) distribution at 10m

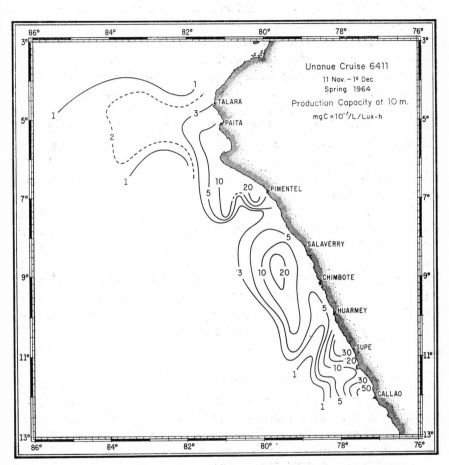

Figure 5b Productivity capacity (mg C × 10⁻⁷/L/Lux-h) distribution at 10 m

The average productivity values, however, were slightly higher due to the scanty data and the fact that two high values were included in the calculation. Below 6° S higher values occurred with an average of 12–98 mg C × 10^{-7}/L/Lux-h. The highest values were found in the upwelling areas off Huarmey and Supe with 69.9 and 37.0 mg C × 10^{-7}/L/Lux-h respectively. Off Cabo Blanco and near 82° W there was an area with 23.7 mg C × 10^{-7}/L/Lux-h of productivity, higher than the summer value for the same area and probably due to enrichment by mixing.

During the winter (Fig. 5), despite intense upwelling, the lowest values occurred for the year with an average of 5.09 mg C × 10^{-7}/L/Lux-h south of 6°S due to lack of enough light. The areas of better productivity were off Salaverry-Chimbote and Pimentel with 18.8 and 10.5 mg C × 10^{-7}/L/ Lux-h, respectively. In the northern region off Talara-Paita, and to the west of 82° W, there was an area with values greater than 3.0 mg C × 10^{-7}/L/ Lux-h. This was due to mixing of equatorial waters running south meeting the northern flowing waters of the Peruvian Coastal Current.

During the spring (Fig. 5), productivity values increased relative to the winter ones due to the increase in stability and light as well as the better provision of nutrients brought up by upwelling. The better concentrations occurred south of 6° S along the coast with an average of 9.73 mg C × 10^{-7}/L/ Lux-h, Pimentel, Supe and Callao with 26.1, 33.7 and 53.3 mg C × 10^{-7}/L/ Lux-h respectively were notably higher. North of 6° S and to the west of 83°W, the area was occupied mostly by surface equatorial waters and its productivity was not as low as expected because of the mixing equatorial and Peruvian Coastal Current Waters.

Distribution of phytoplankton at 10 m

In the summer (Fig. 6), phytoplankton values were higher south of 6° S and near the coast with an average of 265 × 10^3 cells/L. The areas of major phytoplankton productivity coincided very closely with the zones of higher productivity capacity during the summer in the areas off Pimentel, Salaverry and Supe with 2176, 1134 and 2690 × 10^3 cells/L respectively and off Callao with the highest (5538 × 10^3 cells/L), for which, however, there are no data on productivity capacity. The poorer concentrations of phytoplankton occurred in areas north of 6°S and at a considerable distance from the coast.

During the autumn (Fig. 6), the phytoplankton concentrations generally decreased as compared with the summer values. The greater values were

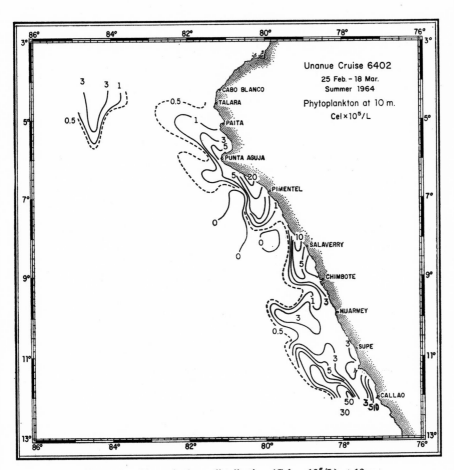

Figure 6a Phytoplankton distribution (Cel $\times$ 10⁵/L) at 10 m

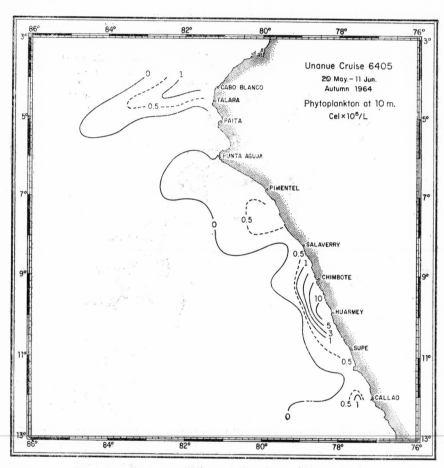

Figure 6b Phytoplankton distribution (Cel $\times$ 10^5/L) at 10 m

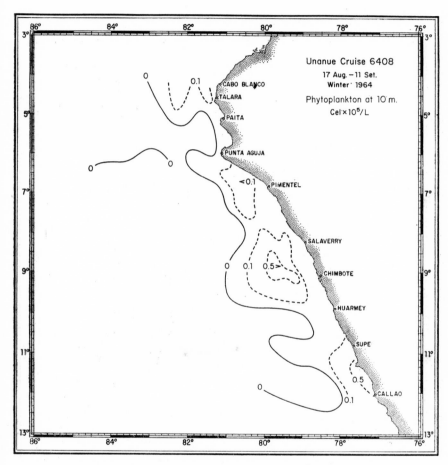

Figure 7a Phytoplankton distribution (Cel $\times$ 10^5/L) at 10 m

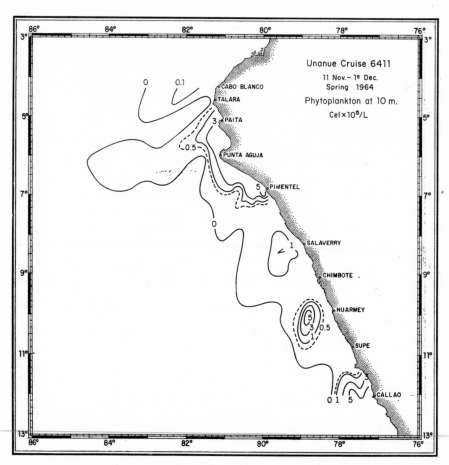

Figure 7b   Phytoplankton distribution (Cel $\times$ 10^5/L) at 10 m

found south of 6° S with an average of 130 × 10³ cells/L. The area off Huarmey was outstanding with 1007 × 10³ cells/L which also coincided with this area of best productivity. North of 6° S, and approximately 82° W in the area off Cabo Blanco, there is a zone of high phytoplankton concentration which also coincides with the area of best productivity north of 6° S.

In the Winter (Fig. 7), phytoplankton concentrations were the least for the year with an average of 18 × 10³ cells/L for the areas south of 6° S. Exceptions were found off Salaverry and Chimbote with 76 and 73 × 10³ cells/L respectively. These areas also coincided with the ones of high productivity during the same season. North of 6° S no phytoplankton concentrations were found except in the area off Cabo Blanco which also coincided with the area of best productivity.

In the spring (Fig. 7), phytoplankton concentrations generally increased with respect to the Winter value due to better light and to the presence of upwelling with its transport of nutrients to the surface. The greater concentrations occurred south of 6° S and near the coast with an average of 96° × 10³ cells/L. Especially noteworthy were the areas off Pimentel (657 × 10³ cells/L), Huarmey (805 × 10³ cells/L) and Callao (766 × 10³ cells/L), which coincided very closely with the best productivity areas. North of 6° S, phytoplankton concentrations were reduced except for the area close to the coast between Paita and Punta Aguja.

Seasonal variations of the production capacity and of phytoplankton

These variations are shown in Figures 8 and 9. Figure 8 shows that the higher values of productivity occurred south of 6° S with maxima in summer and minima in winter, averaging 15.05 and 5.79 × 10⁻⁷ mg C/L/Lux-h, respectively. The best productivity area was between 11 and 12° S with an average of 17.29 × 10⁻⁷ mg C/L/Lux-h. Following were the areas between 9°–10° S, 8°–9° S and 6°–7° S with average values of 14.56, 12.45, and 10.24 × 10⁻⁷ mg C/L/Lux-h respectively.

Figure 9 shows that the phytoplankton distribution was similar to that of the productivity capacity, with the greater concentrations south of 6° S. The maxima occurred also during the summer and the minimal during the winter. The areas of best productivity coincided with those of greatest phytoplankton concentration between 11°–12° S, 9°–10° S and 6°–7° S with averages of 168, 172 and 190 × 10³ cells/L respectively.

The combined effect of vertical turbulence, greater turbidity, and reduced radiation was enough to check productivity during the winter. In the summer,

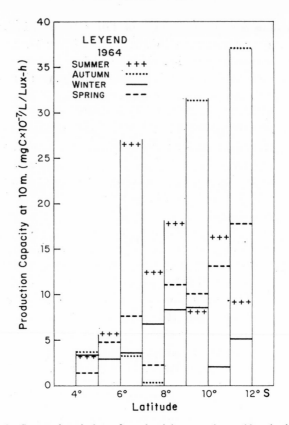

Figure 8 Seasonal variation of productivity capacity at 10 m by latitude

regeneration and provision of nutrients to the euphotic zone as a consequence of better light conditions and upwelling apparently resulted in higher values.

Figure 10 gives the seasonal per cent distribution of phytoplankton species within 60 miles of the coast during 1964. It can be seen that of the 5 dominant species, *Eucampia zoodiacus, Rhizoso lenia delicatula, Thalassiosira subtilis, Skeletonema costatum and Chae toceros sp.* The first 3 made up 50% of the total phytoplankton during the summer. For the Autumn, *Eucampia zoodiacus, Nitzchia pungens* and *Chaetoceros sp.* made up 82%. In the Winter, 62% was made up by *Coscinodiscus sp., Chaetoceros sp.* and *Planktoniella costatum* while during the spring, *Skeletonema costatum, Eucampia zoodiacus* and *Thalas siosira subtilis* accounted for 80% of the

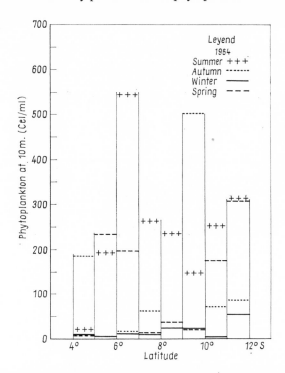

Figure 9   Seasonal variation of phytoplankton (Cel/ml) at 10 m by latitude

total among the dominant species. Three species were present all the year
and along all the coast. (*Eucampia zoodiacus, Skeletonema costatum* and
Rhyzosolenia delicatula.) These had an average count of 108 × 10³ cells/L
per positive station, equivalent to 51 % of the total phytoplankton. The only
genus present all the year was *Chaetoceros.* South of 6° S *Thalassiosira sp.,*
Asterionella japonica and *Thalassionema bacillaris* were also present plus
some dinoflagellates.

Table No. 1 shows the population characteristics of the plankton at 10 m
and by latitude. It is noted that the greater cell concentrations were found
during the summer and the smaller during the winter. Throughout the entire
year, the greatest concentrations are always found south of latitude 6° S.
Dominance (Table 1), is given by the relation of the concentration of the
two most dominant species to the number of cells in the sample (Hulburt,
1963), and is correlated to the population size. It is very difficult to define
and recognize the variations of those species which occur repeatedly with

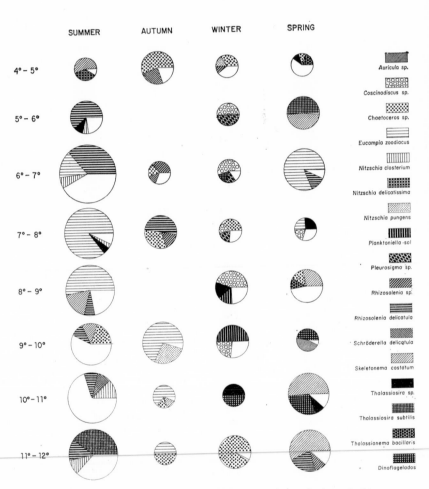

Figure 10 Seasonal variation of main components of phytoplankton (Cel/ml) by latitude

Table 1 Characteristics of the populations

Latitude	Summer			Autumn			Winter			Spring		
	No. of Samples	Average Cell No./ml	Average % Dominance	No. of Samples	Average Cell No./ml	Average % Dominance	No. of Samples	Average Cell No./ml	Average % Dominance	No. of Samples	Average Cell No./ml	Average % Dominance
4°–5° S	4	20	90	1	183	80	7	8	50	3	7	30
5°–6° S	5	194	71	—	—	—	2	2	100	3	231	99
6°–7° S	6	544	53	4	15	54	3	9	69	8	194	93
7°–8° S	9	261	91	2	59	80	5	9	55	5	11	51
8°–9° S	12	229	71	—	—	—	8	21	61	8	38	42
9°–10° S	12	145	32	2	504	95	7	21	73	4	16	94
10°–11° S	10	250	21	1	72	79	1	4	100	8	174	81
11°–12° S	7	312	53	1	38	100	2	55	93	6	310	79

the same relative composition. However, there are some species that are usually found together although not always. Their associated presence has a special value as an indicator of water masses. Composition and association of plankton in the coastal waters off Peru will be discussed in another paper.

The phytoplankton, the production capacity, and their relations with the environment

The annual averages of productivity capacity, phytoplankton, and total productivity along the Peruvian coast in a 60 mile belt are shown in Figure 11. It can be seen that the productivity capacity, as well as the phytoplankton

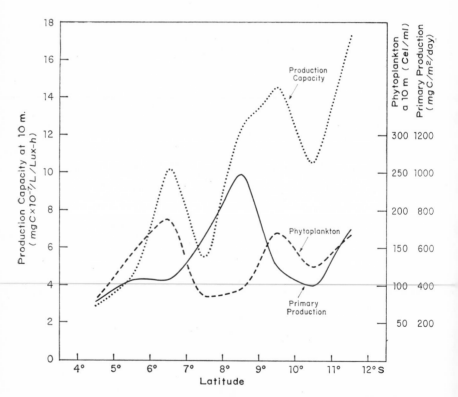

Figure 11   Average variation of productivity capacity, phytoplankton and primary productivity by latitude

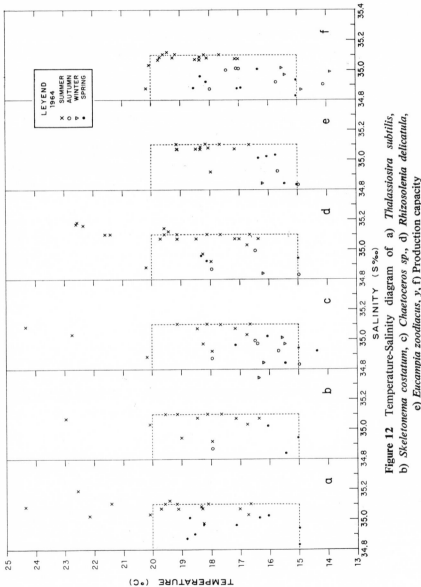

Figure 12 Temperature-Salinity diagram of a) *Thalassiosira subtilis*, b) *Skeletonema costatum*, c) *Chaetoceros sp.*, d) *Rhizosolenia delicatula*, e) *Eucampia zoodiacus*, y, f) Production capacity

distribution, are associated to the primary productivity in the water column (data taken from Guillén and Izaguirre de Rondán, 1969).

The five most abundant phytoplankton species and the stations with the major concentrations of productivity ($>10 \times 10^{-7}$ mg C/L/Lux-h) have been plotted in a T-S diagram at 10 m (Fig. 12). It is seen that both the most abundant species and the higher productivity values are found in waters of the Peruvian Coastal Current with salinities between 35.1 and 34.8⁰/₀₀.

Comparing the figures for primary productivity following the method of Steemann Nielsen (1952) and after Berge (1958), Figures 13 and 14 were made for the 0 and 10 meter levels. These figures show a good correlation between productivity capacity at 10 m compared to primary productivity at 0 m ($r = 0.82$) and between primary productivity in the water column versus primary productivity at 0 m ($r = 0.87$).

From these figures it can be seen that the primary productivity at 0 m or the productivity capacity at 10 m at times could be used as indicaters

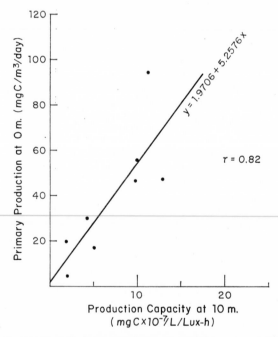

Figure 13   Relationship between primary productivity (mg C/m³/day) 0m and productivity capacity (mg C × 10⁻⁷/L/Lux-h) at 10 m

Table 2 Composition of the phytoplankton (cell/ml) at 10 mts. in equatorial, subtropical, and upwelling waters

	Surface equatorial waters				Surface subtropical waters				Upwelling waters			
	Summer Cruise 6402	Autumn Cruise 6405	Winter Cruise 6408	Spring Cruise 6411	Summer Cruise 6402	Autumn Cruise 6405	Winter Cruise 6408	Spring Cruise 6411	Summer Cruise 6402	Autumn Cruise 6405	Winter Cruise 6408	Spring Cruise 6411
Station No.	37		8	21	98		104	112	48	56	56	73
Diatomeas Centrales												
Chaetoceros sp.			2	2					8			32
Coscinodiscus sp.										13	4	
Dactyliosolen mediterranea									2			
Eucampia zoodiacus									463			598
Planktoniella sol							2					
Rhizosolenia delicatula								4	4			
Rhizosolenia sp.	2								4	21		
Schroderella delicatula								4	4	8		21
Skeletonema costatum									8			4
Thalassiosira subtilis								8				
Thalassiosira sp.									30			
Diatomeas Pennales												
Asterionella japonica								4	11			
Auricula sp.									2			
Nitzschia delicatissima									19			
Nitzschia closterium									2			
Nitzschia pungens												
Thalassionema bacillaris				2	8			2				
Dinoflagelados								4	2		4	2

of the total productivity. That is to say that when the phyto-plankton is physiologically active and abundant at 0 or 10 m it follows through for the euphotic zone. However, the productivity in the euphotic zone is not always accompanied by high productivity at the 0 and 10 m levels.

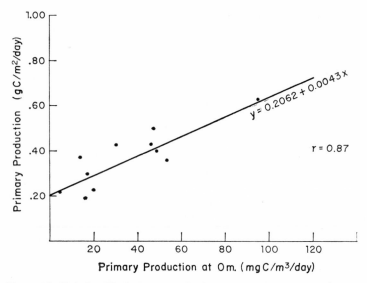

Figure 14   Relationship between total primary productivity (g C/m²/day) and primary productivity (mg C/m³/day) at 0 m

To show the distribution and composition of the plankton at 10 m with their respective water masses, representative stations were chosen from 4 cruises in 1964 (Table 2). This indicates that the great variety and concentration of plankton corresponds to the upwelling areas, followed by the Subtropical and Equatorial surface masses.

CONCLUSIONS

Study of phytoplankton and primary productivity along the Peruvian coast, between the 4 and 12° S during 1964 shows the following:

1) The areas of major productivity were associated with the major phytoplankton concentrations and were found south of the 6° S along and near the coast; these areas benefit in nutrients brought up by upwelling.

2) High productivity areas were also found off Pimentel, Salaverry-Huarmey and Callao. They shifted slightly with the seasons due to variations in the intensity of upwelling and also due to circulation processes which modify the distribution of surface water masses.

3) The most abundant phytoplankton species as well as the higher productivity values were found in waters of the Peruvian Coastal Current with salinities between 35.1 and 34.8$^0/_{00}$.

4) The only genus found during all the year was *Chaetoceros*. South of 6° S, *Coscinodiscus sp.*, *Eueampia zoodiacus*, *Skeletonema costatum*, *Thalassiosira sp.*, *Asterionella japonica*, *Nitzschia pungeus* and *Thalassionema bacillaris* and some dinoflagellates were also found.

5) Productivity capacity at 10 m and primary productivity at 0 m can sometimes be a good measure of the total productivity.

References

BERGE, G. (1958). The Primary Production in the Norwegian Sea, June 1954, Measured by an adapted 14C Technique. *Rapp. Cons. Explor. Mar.* **144**, 85–91.

GUILLÉN, O. and IZAGUIRRE DE RONDÁN, R. (1968). Producción de las Aguas Costeras del Perú en el año 1964. *Inst. Mar Perú* **1**, 7, 394–76.

HULBURT, E. M. (1963). The diversity of phytoplanktonic populations in oceanic, coastal and estuarine regions. *J. Mar. Res.* **21**, 81–93.

ROJAS DE MENDIOLA, B. (1958). Breve estudio sobre la variación cualitativa anual del plancton superficial de la Bahia de Chimbote. *Bol. Cia. Adm. Guano.* **34**, 12, 7–17.

ROJAS DE MENDIOLA, B. (1963a). Estudios preliminares sobre la distribución del fitoplancton en Noviembre de 1962, en el área del Callao-Cabo Blanco. *Inf. Int. Inst. de Inv. de los Recurs. Mar.*, 34, 10 pp (Manuscrito).

ROJAS DE MENDIOLA, B. (1963b). Análisis cuantitativos y cualitativos del fitoplancton en Enero de 1963, en el área des pesca Supe-Cerro Azul. *Inf. Int. Inst. de Inv. de los Recurs. Mar.*, 38, 10 pp (Manuscrito).

ROJAS DE MENDIOLA, B. (1963c). Distribución del fitoplancton en Agosto de 1961, en el área de Callao-Chimbote. *Inf. Int. Inst. de Inv. de los Recurs. Mar.*, 88, 12 pp (Manuscrito).

STEEMANN NIELSEN, E. (1952). The use of radio-active carbon (14C) for measuring organic production in the sea. *J. Cons. Explor. Mer.* **18**, 117–40.

STRICKLAND, J. D. H., EPPLAY, R. W. y ROJAS DE MENDIOLA, B. (1969). Poblaciones de fitoplancton, nutrientes y fotosintesis en aguas costeras peruanas. *Inst. Mar. Perú.* **2**, 1, 4–12.

ZUTA, S. y GUILLEN, O. (1964). "Condiciones Oceanográficas frente a las costas del Perú en 1964". (Manuscrito.)

The "El Niño" phenomenon in 1965 and its relations with the productivity in coastal Peruvian waters

OSCAR GUILLÉN

Instituto Del Mar
Lima, Peru

Resumen

Los datos para el presente travajo fueron tomados del crucero B.A.P. "Unanue" 6504. La concentracion de clorofila "a" fue usada como un indice de la cosecha estable del fitoplancton, la produccion del fitoplancton fur medida por la técnica del C^{14} bajo condiciones naturales de luz y a temperaturas superficiales.

La distribución de la concentración del fitoplancton y su tasa fotosintética en la superficie del mar, así como también su producción primaria en la columna del agua y su relacion con las aguas superficiales son mostradas.

La distribución de la productivadad a lo largo de la costa peruana estuvo intimamente asociada a la circulación de las aquas. El flujo de las aquas ecuatoriales superficiales con temperaturas de 27–24°C, salinidades de 34.8–33.8°/oo y bajísimo contenido de fosfatos habían avanzado hacia el SE en forma de una lengua hasta frente a Supe. Este mismo flujo estuvo asociado con las bajisimas concentraciones de fitoplancton y produccion, hallandose las grandes concentraciones nuy cerca de la costa, en aguas de la Corriente Costera Peruana con temperaturas de 22–16°C y salinidades menores de 35.1°/oo. Estos altos valores durante el fenomeno "El Nino" fueron más bajas que el promedio para estas mismas áreas durante años normales.

El afloramiento a lo largo de la costa reducido, siendo más intenso frente a Pisco, en donde se halló una producción de 0.80 gr C/m²/día, equivalente a una producción anual de 290 gr C/m².

Las más grandes concentraciones de anchoveta se hallaron en el área de mayor productividad frente a salaverry con una producion de 1.56 gr C/m²/día.

INTRODUCTION

According to Wyrtki (1966) the surface temperatures off Peru and Ecuador have an annual variation of 5 to 7°C. Although this variation is mainly due to local warming in this area, farther north it is intensified by the southward shift of warm tropical surface waters from December to February.

In some abnormal years, the warm waters reach much further south, reducing the coastal cold water areas. This situation is known as the "El Nino" phenomenon and has catastrophic effects on the guano bird population, on the distribution of pelagic fish, and on the coastal climate. It has been recorded and described for 1891 (Schott, 1931), 1925 (Murphy, 1926), 1941 (Lobell, 1942 and Schweigger, 1942), 1953 (Wooster and Jennings, 1955 and Posner, 1957), 1957–58 (Wooster, 1960, Bjerknes, 1967) and 1965 (Guillen 1967).

As the relation of this phenomenon with productivity has not been duly studied yet, this report is presented as a first contribution toward the understanding of this phenomenon.

METHODS AND OBSERVATIONS

This report utilizes data that were obtained during cruise 6504 (April 1965) by B.A.P. "Unanue". Chemical analyses were made following the techniques and modifications of Strickland and Parson (1965). Carbon fixation rate was measured by the radio carbon (C^{14}) method as described by Steemann Nielsen (1952). The samples were taken with Van Dorn bottles at depths corresponding to 100, 28, 10 and 2.8% of surface light intensity. The radioactivity of the C^{14} was of 4 µc and the samples were incubated at surface temperature under natural light from sunrise to LAN or from LAN to sunset. Chlorophyll "a" was extracted with acetone following the technique described by Strickland and Parson (1965) and was used as an index of phytoplankton standing crop.

RESULTS AND DISCUSSION

Distribution of the physical-chemical properties of sea water

The oceanographic conditions off Peru during the cruise have been described by Guillén (1967). The equatorial surface waters with temperatures between 27 and 24°C, salinities from 34.8 to 33.8⁰/₀₀ and very low phosphate content, advanced to the SE as a tongue approximately 30 m deep, extending as far south as Supe; below 11°S latitude temperatures descreased while salinity increased due to the presence of the Peruvian coastal current with temperatures 22–16°C and salinities less than 35.1⁰/₀₀.

Nutrients, primary production, and chlorophyll "a"

Only phosphate analyses were made. Surface phosphates were lower than usually found in this area (Wooster and Cromwell, 1958, Guillén 1964).

Primary production values were greater than 0.50 g C/m²/day along the Peruvian coast with the highest value off Salaverry (1.56 g C/m²/day). Highest concentration of anchoveta were also found in this area. The area of most intense upwelling was off Pisco with a production of 0.80 g C/m²/day. On the other hand, the lowest production occurred in areas of the equatorial surface waters with values below 0.05 g C/m²/day. The surface subtropical waters, found south of 11°S, were low in production, less than 0.10 g C/m²/day, and had greatly influenced the near shores areas of Callao.

Figures 1 and 2 show phytoplankton concentrations (Zuta and Guillén 1970) and their photosynthetic rates for the surface. The distribution of values is similar to that for total production; that is, higher values than 0.6 µg/L of chlorophyll "a" and 5.0 mg C/m³/day for primary production were found along the coast. The lower values, less than 0.2 µg/L and 1.0 mg C/m³/day for chlorophyll and primary production, occurred in areas occupied by the equatorial surface waters.

Productivity and its relations with the 1965 "El Niño" phenomenon

The distributions of salinity, temperature, chlorophyll "a", and primary production show that the productivity of coastal waters off Peru was associated with the mass circulation. The higher productivity values were found along and near the coast. The lowest productivity values were found in areas occupied by surface equatorial waters. The higher values during the El Niño phenomenon were still lower than the average value found in these same areas during normal years.

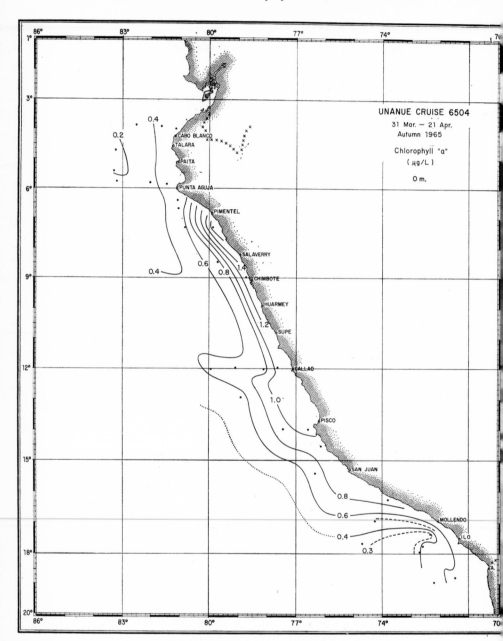

Figure 1 Chlorophyll a (μg/L) distribution at 0 m

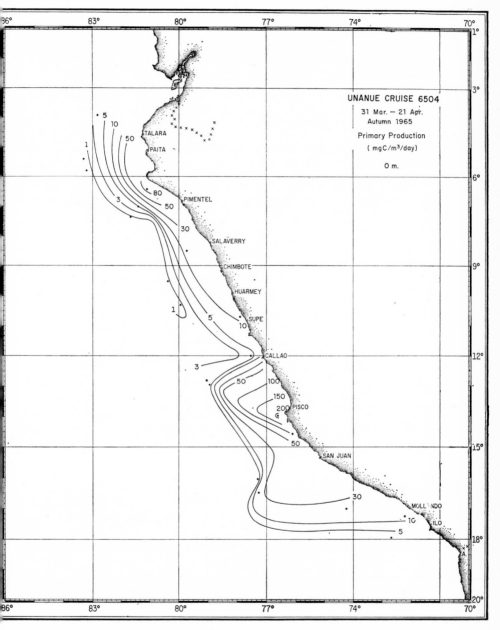

Figure 2 Primary Production (mg C/m³/day) at 0 m

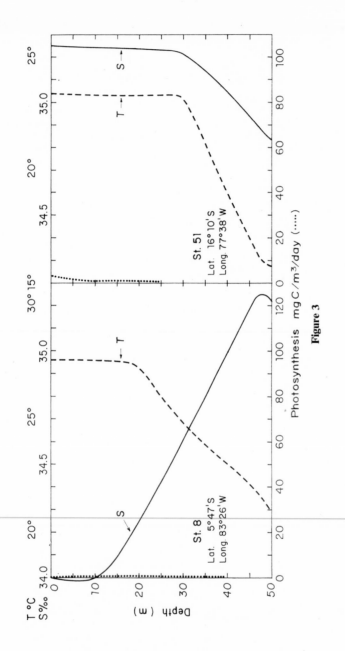

Figure 3

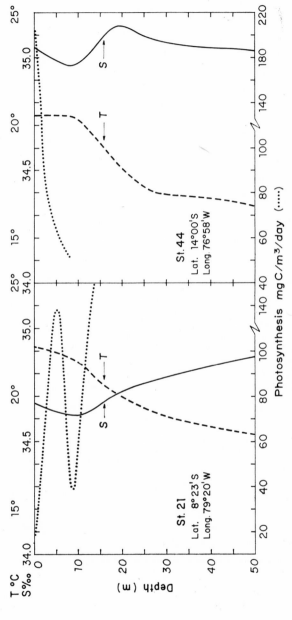

Figure 3 Profiles of photosynthetic rates with temperatures and salinities for stations 8, 51, 21 and 44

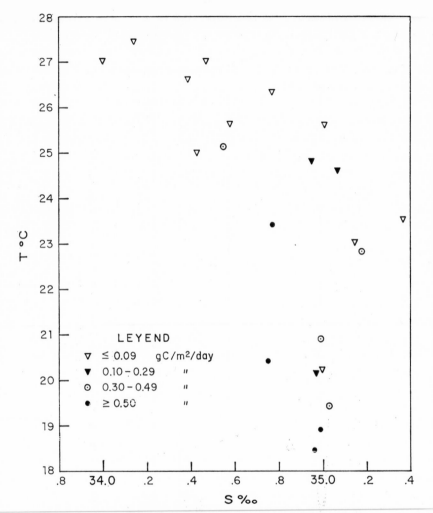

Figure 4 Temperature-salinity diagram for average values at 0–20 m and corresponding primary production

In order to compare the productivity of the surface waters, the following stations were selected according to type water masses.

Station 8 (Fig. 3) was in the area occupied by the equatorial surface waters where the productivity was very low (0.01 g C/m²/day) due probably to lack of nutrients. Station 50 was located in surface subtropical waters with the same low values, probably for the same reason. Station 44, located in the area of most intense upwelling off Pisco, had a productivity of 0.80 g C/m²/day. Station 21 off Salaverry had higher productivity (1.56 g C/m²/day) along with very high oxygen values (>6.0 ml/L) and very low phosphates (<0.5 μg-at/L). In this same area the highest anchoveta concentrations were also found.

Figure 4 helps to give a better understanding of the relation between the different water masses at the surface and their productivity. It shows that average temperature and salinity of the 0–20 m layer and their corresponding total production values for each station. The higher production values were found in waters of the Peruvian coastal current, thus agreeing with Guillén and Izaguiree de Rondan (1968) and with Strickland *et al.* (1969). The detailed relationships between the 1965 "El Niño", the water masses, and the surface circulation is considered by Zuta and Guillén (1970).

CONCLUSIONS

The distribution of productivity along the Peruvian coast is associated with water circulation. The flow of the surface equatorial waters with temperatures and salinities between 27–24°C, and 34.8–33.8°/₀₀, respectively, advanced to the SE in a tongue reaching as far south as Supe. While very low concentrations of phytoplankton and very low productivity were characteristic of this flow, the higher phytoplankton concentrations and higher productivity occurred very near to the coast in waters of the Peruvian Coastal Current with temperatures 22–16°C and salinities less than 35.1°/₀₀. These higher values during the El Niño phenomenon were still lower than the average value found in these same areas during normal years. General coastal upwelling was reduced with the greater intensity being off Pisco where a productivity of 0.80 g C/m²/day, equivalent to 290 g C/m²/year was found. Greater concentrations of anchoveta were found in the areas of higher productivity off Salaverry with a production of 1.56 g C/m²/day.

References

BJERKNES, J. (1967). Survey of "El Niño" 1957–58 in its relation to tropical Pacific Meteorology. *Inter. Amer. Trop. Tuna Comm. Bull.* **12**, 2, 1–42.

GUILLEN, O. (1964a). Distribución del contenido de fosfatos en la región de la Corriente Peruana. *Inf. Inst. de Inv. de los Rec. Mar.* **28**, 1–15.

GUILLEN, O. (1964b). Distribución y variación anual de fosfatos y oxigeno disuelto en la región maritima del Callao. (Feb. 1961–Enero 1962). *Inf. Inst. Inv. Recurs. Mar.* **28**, 1–16.

GUILLEN, O. (1964c). Distribución y variación anual de fosfatos y oxigeno disuelto en la región maritima del Callao durante el ano 1962. *Inf. Inst. Inv. Recurs. Mar.* **28**, 1–7.

GUILLEN, O. (1967). Anomalies in the waters off the Peruvian cost during March and April 1965. *Stud. Trop. Oceanogr. Miami.* **5**, 452–465.

GUILLEN, O. and IZAGUIRRE DE RONDAN, R. (1968). Produccion Primaria de las Aguas Costeras del Peru en el ano 1964. *Inst. Mar. Peru. Bol.* **1**, 7, 349–376.

LOBELL, M. J. (1942). Some observations on the Peruvian Coastal Current. *Trans. Amer. Geophys. Union.* **2**, 332–336.

MURPHY, R. C. (1926). Oceanic and climatic phenomena along the west coast of south America during 1925. *Geogr. Rev.* **16**, 26–54.

POSNER, G. S. (1957). The Peru Current. *Bull. Bingham Oceanogr. Coll.* **16**, 2, 106–155.

SCHOTT, G. (1931). Der Peru-Strom und seine nördlichen Nachbargebiete in normaler und abnormaler Ausbildung. *Ann. Hydrographie u. Marit. Meterorologie.* **59**, 161–253.

SCHWEIGGER, E. H. (1942). Las irregularidades de la Corriente de Humboldt en los años 1925 a 1941, una tentative explicación. *Bol. Soc. Adm. Guano.* **18**, 27–42.

STEEMANN NIELSEN, E. (1952). The use of radio-active carbon (C^{14}) for measuring organic production in the sea. *J. Cons. Explor. Mar.* **18**, 117–140.

STRICKLAND, J. D. H. and PARSONS, T. R. (1965). A Manual of Sea Water Analysis. *Research Board of Canada. Bull.* **125**, 1–203.

STRICKLAND, J. D. H., EPPLEY, R. W. and ROJAS DE MENDIOLA, B. (1969). Poblaciones de fitoplancton, nutrientes y fotosintesis en aguas costeras peruanas. *Inst. Mar. Perú.* **2**, 1, 4–12.

WOOSTER, W. S. (1960). "El Niño". *Calif. Coop. Ocean. Fish. Invest. Rept.* **7**, 43–45.

WOOSTER, W. S. and JENNINGS, F. (1955). Exploratory oceanographic observations in the eastern tropical Pacific January to March 1953. *Calif. Fish. Game.* **41**, 1, 79–90.

WOOSTER, W. S. and CROMWELL, T. (1958). An oceanographic description of the eastern tropical Pacific. *Bull. Scripps Instn. Oceanogr. Univ. Calif.* **7**, 169–282.

WYRTKI, K. (1966a) Oceanography of the Eastern Equatorial Pacific Ocean. *Oceanogr. Mar. Biol. Ann. Rev.* **4**, 33–68.

ZUTA, S. and GUILLEN, O. (1970) On the Oceanography of the Peru Coastal Waters. *Inst. Mar. Perú.* (En prensa).

Determination of microbial biomass in deep ocean profiles*

OSMUND HOLM-HANSEN

Institute of Marine Resources
University of California
La Jolla, California

Resumen

A maior parte do carbono orgânico do Oceano é encontrada nos compostos orgânicos dissolvidos e na fração detrital (não viva) particulada.

Para compreender o ciclo geral de energia no Oceano, é importante ter uma estimative da distribuição de células vivas em águas profundas, sua atividade metabólica e seu significado em cadeias tróficas de profundidade. Nêste trabalho são discutidos os vários métodos de medida da biomassa microbial ou atividade metabólica com especial atenção ao suo do método de medida de ATP (trifosfato de adenosina). Os resultados de perfis oceânicos indicam que, primeiro, a concentração de ATP na zona eufótica é alta, e corresponde a estimativas de biomassa as quais estão em excelente concordância com aquelas baseadas em medidas de clorofila, número de células e volumes. Segundo, a concentração de ATP diminui ràpidamente entre 100 e 200 metros de tal maneira que nesta profundidade a concentração de ATP indica uma biomassa de cêrca de 1.0 microgramas de carbono orgânico/litro. Terceiro, entre 200 e 3.000 metros a diminuição de concentração de ATP é gradual. A 3.000 metros, as estimativas representam cêrca de 0.1 micrograma de carbono/litro.

*This research was supported in full by the U.S. Atomic Energy Commission, Contract No. AT(11-1) GEN 10, P.A. 20.

INTRODUCTION

Our primary interest for determination of the quantitative distribution of microbial cells in deep ocean water is related to food-chain dynamics. The only significant input of energy-rich organic material into the oceans is via the photosynthetic reduction of carbon dioxide, which occurs almost entirely in the upper 100 meters of the water column. We know, however, that there is a wide diversity of microbial and animals species at all depths in the oceans. There is not much information on the quantitative abundance of bathypelagic species, but it appears that the distribution of net plankton biomass below 1000 meters decreases in an exponential fashion (Vinogradov 1962a). The bottom sediments, even in the trenches at depths greater than 10,000 m, also contain a wide variety of bacteria and animals (Zobell and Hittle 1969; Sanders and Hessler 1969; Vinogradova 1962). The base of the food chains for all this bathypelagic life is not known, but the following hypotheses have been suggested. (1) Vinogradov (1962b) has postulated that there is a ladder-like series of migrating zooplankton populations which results in an active transfer of food materials from the euphotic zone to all depths in the water column. (2) Deep ocean water usually contains between 5 to 10 µg organic carbon in particulate material which passes through a 150 µ mesh nylon net but which is retained by filters with pore sizes of about 0.5 µ (Menzel and Ryther 1968; Holm–Hansen 1969). This material, which has a carbon/nitrogen ratio similar to the particulate material in the euphotic zone (Holm–Hansen 1969; Hobson and Menzel 1969), might serve as food for filter-feeding zooplankters. In addition to this 0.5 to 150 µ particulate material, there may be significant amounts of larger particles (such as fecal pellets and molts) descending in the water column. As there seems to be an equilibrium between dissolved organic compounds and the particulate material, the consumption of particulate matter by filterfeeding animals would not necessarily result in decreasing concentrations of particulate material with depth. (3) The concentration of dissolved organic compounds in the aphotic zone is also fairly uniform with depth, the range being about 0.2 to 0.8 mg organic carbon per liter with the average value being close to 0.5 mg carbon/liter (Menzel and Ryther 1968). This reservoir of dissolved organic carbon, together with the detrital particulate fraction, accounts for well over 98 % of all the organic carbon in the oceans, the remaining carbon being found in living plant or animal cells. These reservoirs of reduced carbon might support the growth of a heterotrophic microbial population, which in turn may serve as food for filter-feeding zooplankton species.

The first two hypotheses above do not necessarily implicate a heterotrophically growing microbial population as being important in the food chains occurring in deep ocean water. As there is no evidence that macroscopic zooplankton can exist solely by assimilation of dissolved organic nutrients, the third hypothesis would require the existence of a microbial population in deep water. The detection and quantitative estimation of such a population is described in the following sections.

METHODS TO ESTIMATE BIOMASS

The estimation of biomass in the euphotic zone can be obtained from chlorophyll data (Jørgensen 1966), direct microscopic analysis (Strickland *et al.* 1969), or by growth rate measurements (Epply 1968). The problem of estimating biomass or the activity of living cells in deep ocean water is much more difficult. The following list includes most of the commonly used methods for such estimations.

1) The measurement of such chemical entities as organic carbon, nitrogen, or nucleic acids might be useful for certain eutrophic waters, but is not useful in deep ocean water. As the detrital (non-living particulate material) fraction contains these same elements or compounds, such analyses give little information about biomass (Holm–Hansen 1969; Holm–Hansen *et al.* 1968).

2) The classical technique of counting bacterial colonies on agar surfaces is not very informative as it tells you nothing about non-bacterial cells and is also highly selective for certain species. Colony counts will generally amount to a few percent or less of the total bacterial count in the sample (Jannasch and Jones 1959).

3) A direct microscopic analysis for cell numbers and cell volumes can yield a reasonable estimate of biomass in samples from the euphotic zone (Holm–Hansen 1969). It is difficult, however, to differentiate between live and dead cells, and the technique becomes most laborious and time consuming for deep-water samples where most of the cells seen are less than 5 microns in diameter (Hamilton *et al.* 1968). Differentiating between small cells and detrital material is so difficult that I tend to discount this method for deep-water estimates of biomass.

4) It is possible to obtain the number and size of the suspended materials in unconcentrated water samples by use of an electronic particle counter. This technique counts all particles and does not discriminate between living cells and detrital material. A method has been described, however, which permits one to estimate the amount of living material by repeated analyses of the size distributions in a sample which is incubated to permit growth of any living cells (Strickland and Parsons 1968). I do not think, however, that this technique will prove very useful in the estimation of biomass in deep ocean water.

5) Many investigators (Vaccaro 1969; Wright and Hobbie 1965) have attempted to obtain data on the heterotrophic potential in water samples by studying the rate of assimilation of radio-labelled organic substrates. Although such studies are valuable for some types of studies, I do not think they will yield very meaningful estimates of total biomass or activity due to such factors as substrate specificity, concentrations of naturally-occurring substrates, and enzyme induction lag periods.

6) Pomeroy and Johannes (1968) have concentrated the particulate material from about 300 liters of ocean water to about 25 ml by the reversed membrane filter technique and then measured oxygen assimilation in the concentrate by use of an oxygen electrode. Although these techniques have yielded interesting data on respiratory rates in the euphotic zone, the sensitivity of the method makes it very marginal for studies of water samples from below a few hundred meters.

7) Recently Packard (1969) has described a method for estimation of respiratory rates based on the maximal rate of electron transport as determined by the reduction of a tetrazolium dye compound. He has estimated the rate of oxygen utilization from the surface down to 5800 meters in tropical Pacific waters. This method, which has the sensitivity to estimate the respiratory rate at any depth in the ocean, has not as yet been subjected to a critical laboratory examination, but the field data I have seen indicate that the method may have great promise. It is to be noted that Packard does not claim that his measurements necessarily indicate respiration rates *in situ*, but that they do indicate the maximal potential rate of electron transfer in the cellular material from any particular water depth.

8) Another approach to the problem of estimating biomass is to quantitatively analyze for some cellular constituent which meets the following criteria. (a) The constituent to be determined must exist in all live cells at

fairly uniform concentrations. (b) It must not be found in dead cells or in the detrital fraction. (c) As the concentration of living cells in deep water is very low, there must be a very sensitive analytical method for the detection of the cellular constituent in sub-microgram quantities. The only cellular metabolite that we have found so far that satisfied these criteria is adenosine triphosphate (ATP). The rest of this paper deals with a description of the ATP method for biomass estimations and some field data to indicate its application.

OUTLINE OF ATP METHOD

The quantitative analysis for ATP in sub-microgram quantities depends upon the measurement of the light emitted when ATP is added to an enzyme preparation containing luciferase and luciferin obtained from ground up firefly (*Photinus pyralis*) lanterns. The reactants and products of this bio-luminescent reaction are:.

$$\text{luciferin (reduced)} + \text{ATP} + \text{O}_2 \xrightarrow[\text{Mg}^{++}]{\text{luciferase}} \text{luciferin (Oxidized)}$$
$$+ \text{AMP} + \text{P—P} + \text{H}_2\text{O} + \text{hv.}$$

The details and mechanisms of this reaction have been studied intensively by McElroy and colleagues (McElroy and Strehler 1949; Seliger and McElroy 1960). When all reactants except ATP are in excess, the rate of the reactions is limited by the ATP concentration and there is one photon of light emitted (peak wavelength at about 560 mμ) for each molecule of ATP which is hydrolyzed to adenosine monophosphate (AMP) and pyrophosphate (P—P). To determine the ATP content of any sample therefore, all one has to do is measure the amount of light emitted over some set time interval. Internal standards of ATP are generally employed to equate light readings to ATP concentrations, so it is not essential to know the absolute efficiency of the light-detecting device. If one is working with fairly high levels of ATP (mg quantities), the measurement of the emitted light is a relatively simple matter and can be done with various commercial spectro-photometers and fluorometers. In deep-ocean samples, however, one is generally working with quantities of ATP between 10^{-2} to 10^{-4} microgram, which necessitates the use of a very sensitive light-detecting device. The instrument we have developed (Holm–Hansen and Booth 1966) utilizes a 14-stage photomultiplier tube operating at slightly over $-2000\,\text{V}$. The anode signal is amplified and the amount of light is integrated over a one to two minute period. The excellent review by Strehler (1968) discusses many alternate ways in which the emitted light may be detected and measured. Using a crude enzyme preparation which shows some background emission of light due to transphosphorylase reactions, our lower limit for measurement of ATP is about 1×10^{-5} µg. By use of larger volumes of sample and purified enzyme preparation, it should be possible to increase the sensitivity of our measurement to about 10^{-7} µg ATP.

If one is interested in the quantitative distribution of microbial life in deep ocean water, it is essential to use sampling devices which give you little or no contamination. In most of my early work in deep water I used disposable sterile plastic bags mounted on a Niskin aluminium frame. Samples from these bags, which open at the desired depth and are automatically sealed when filled, should not have any contamination from the water through which the bag passes. These bags can not be used, however, for particulate carbon or nitrogen analyses, as the plastic bags apparently give off some particulate material containing these elements. In recent cruises I have also used alcohol-scrubbed Van Dorn bottles (8.0 liter, constructed of polyvinylchloride) and large (100 to 200 liter) stainless steel samplers. Water samples from these sampling devices which flush on the way down yield comparable ATP data as samples collected in Niskin bags.

As cellular ATP levels are influenced by a change in the environmental conditions, all samples are filtered at 1/3rd atm. as quickly as possible through 47 mm membrane filters (pore size 0.45 µ). The sample volume to be filtered generally is between 0.5 liter for water from the euphotic zone to 3 or 4 liters from water below 2000 m. As soon as the filtration is complete, the filter with all the particulate material is quickly immersed in 5.0 ml boiling Tris buffer (tris[hydroxymethyl]aminomethane; 0.02 M, pH 7.75). After heating at 100°C for five minutes to inactivate all enzymes and extract the cellular ATP, the samples are frozen at −20°C until time of analysis. Samples have been stored for periods up to four months with little or no detectable decrease in ATP concentration. After the samples have been thawed and taken to room temperature, an aliquot is pipetted into the enzyme preparation described above and the ATP content obtained by measurement of the emitted light.

In order to relate ATP concentrations to biomass, I have studied the relation between cellular organic carbon and adenosine triphosphate in a variety of marine and fresh water bacteria and algae under many environmental and physiological conditions. The results indicate that the ATP levels in all the species investigated are fairly uniform and average at about 0.4% of the total cellular organic carbon value. The carbon values have all been obtained by infra-red absorption analysis of CO_2 after wet oxidation of the organic material (Holm–Hansen *et al.* 1967). To convert ATP concentrations to a biomass criterion such as organic carbon, one therefore multiplies the ATP value by 250. The ATP content of the marine copepod *Calanus helgolandicus* has also been investigated and the ATP concentration as a percent of the organic carbon (0.3%) is quite similar to the microbial cells mentioned above.

ATP PROFILES IN THE OCEAN

Figure 1 shows a representative depth profile for ATP down to 3400 m in the eastern Pacific Ocean. The ATP values are always relatively high (50 to 500 mμg ATP/l) in the upper 100 meters of the water column, and then decrease rapidly to values generally between 5 to 20 mμg/l at 200 m. Below 200 m there is a slow, continual decrease in ATP concentrations so that at a few thousand meters depth values are in the range of 0.5 to 2.0 mμg ATP/l. This is a reasonable value for ATP in deep ocean water, as 1 mμg ATP indicates a biomass containing 0.25 μg organic carbon, whereas the total particulate organic carbon at such depths is between 5 to 10 μg/l.

Figure 2 shows the total particulate organic carbon and the organic carbon content of living cells, as calculated from ATP data, in a detailed profile to 600 m off the coast of southern California. The extremely rapid decrease in ATP concentration between 100 and 200 m is even more marked in this profile than that shown in Figure 1. It is seen from Figure 2 that the organic carbon associated with living cells is between 54 and 73% of the total particulate carbon in the upper 100 m, and only 4% at 600 m. The biomass as indicated by ATP analyses in this profile agreed very well with biomass estimates based on chlorophyll values or by direct microscopic observation of cell numbers and volumes (Holm–Hansen 1969).

DISCUSSION

By use of ATP analyses as described above, it is now possible to estimate the biomass of the total microbial standing stock in deep ocean water. This method gives one the total biomass only, and gives no information about

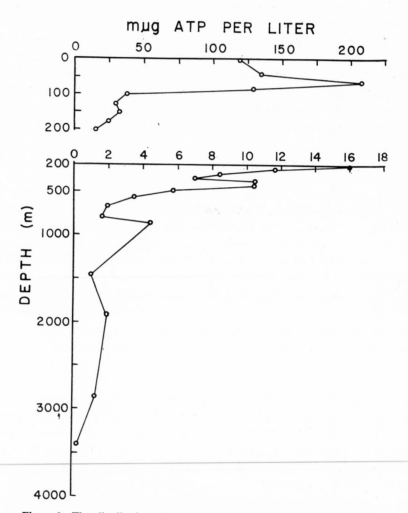

Figure 1 The distribution of adenosine triphosphate with depth in the eastern Pacific Ocean (31° 45′ N, 120° 30′ W). Note that different scales are used for the intervals 0 to 200 m and 200 to 3500 m

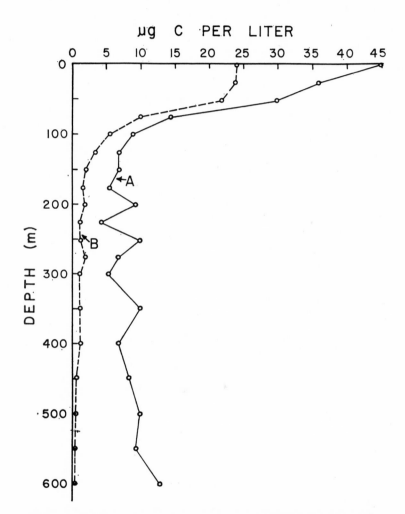

Figure 2   Distribution with depth of total particulate organic carbon (line A) and organic carbon in living cells, as estimated by ATP analyses (line B). Station position was 32° 41′ N, 117° 35′ W

the number or kinds of cells present. The only method available at the present time to give information on the types of cells in any sample is by direct microscopic examination. This, however, is particularly difficult and tedious in deep ocean water, where the amount of living material is very small compared to the detrital fraction. There are no methods available to permit any significant physical fractionation between detrital material and living cells.

Knowledge of the distribution of living cells in the water column is important not only for questions of food-chain dynamics as discussed above, but is also of importance in understanding the distribution and re-cycling of nutrients in the sea. An example of the necessity of knowing the biological activity with depth was afforded by Dr. Postma's talk which dealt with a model system for phosphate distribution in the oceans. So far I have not attempted to extrapolate from ATP values to any metabolic rate such as respiration. Considering the stability of environmental factors in deep ocean water and the fairly uniform concentrations of dissolved and particulate organic carbon, it is possible that ATP values may also be used to estimate metabolic rates such as oxygen uptake or carbon dioxide release in deep water. This extrapolation, however, demands considerable experimental work in the laboratory on the correlation between growth rates, nutrient supply, temperature, and ATP. I am currently working on these related problems to see if ATP values can be used to indicate metabolic turnover rates as well as biomass estimates.

References

EPPLEY, R. W. (1968). An incubation method for estimating the carbon content of phytoplankton in natural samples. *Limnol. Oceanog.*, **13**, 574–582.

HAMILTON, R. D., HOLM–HANSEN, O. and STRICKLAND, J. D. H. (1968). Notes on the occurrence of living microscopic organisms in deep water. *Deep-Sea Res.*, **15**, 415–422.

HOBSON, L. A. and MENZEL, D. W. (1969). The distribution and chemical composition of organic particulate material in the sea and sediments off the east coast of South America. *Limnol. Oceanog.*, **14**, 159–163.

HOLM–HANSEN, O. and BOOTH, C. R. (1966). The measurement of adenosine triphosphate in the ocean and its ecological significance. *Limnol. Oceanog.*, **11**, 510–519.

HOLM–HANSEN, O., COOMBS, J., VOLCANI, B. E. and WILLIAMS, P. M. (1967). Quantitative micro-determination of lipid carbon in microorganisms. *Anal. Biochem.* **19**, 561–568.

HOLM–HANSEN, O., SUTCLIFFE, W. H. JR., and SHARP, J. (1968). Measurement of deoxyribonucleic acid in the ocean and its ecological significance. *Limnol. Oceanog.*, **13**, 507–514.

HOLM–HANSEN O. (1969) Determination of microbial biomass in ocean profiles. *Limnol. Oceanog.*, **14**, 740–747.

JANNASCH, H. W. and JONES, G. E. (1959). Bacterial populations in sea water as determined by different methods of enumeration. *Limnol. Oceanog.*, **4**, 128–139.

JØRGENSEN, C. B. (1966). *Biology of Suspension Feeding.* Pergamon Press. N. Y. 357 p.

McELROY, W. D. and STREHLER, B. L. (1949). Factors influencing the response of the bioluminescent reaction to adenosine triphosphate. *Arch. Biochem.*, **22**, 420–433.

MENZEL, D. W. and RYTHER, J. H. (1968). Organic carbon and the oxygen minimum in the South Atlantic Ocean. *Deep-Sea Res.*, **15**, 327–337.

PACKARD, T. T. (1969). The estimation of the oxygen utilization rate in seawater from the activity of the respiratory electron transport system in plankton. Ph. D. dissertation, University of Washington, Seattle, Washington.

POMEROY, L. R. and JOHANNES, R. E. 1968. Occurrence and respiration of ultraplankton in the upper 500 meters of the ocean. *Deep-Sea Res.*, **15**, 381–391.

SANDERS, H. L. and HESSLER, R. R. (1969) Ecology of the deep-sea benthos. *Science*, **163**, 1419–1424.

SELIGER, H. H. and McELROY, W. D. (1960). Spectral emission and quantum yield of firefly bioluminescence. *Arch. Biochem. Biophys.*, **88**, 136–141.

STREHLER, B. L. (1968). Bioluminescence assay: principles and practice. *In: Methods of Biochemical Analysis* (D. Glick, ed., Interscience Publishers, N. Y.) **16**, 99–181.

STRICKLAND, J. D. H. and PARSONS, T. R. (1968). *A Practical Handbook of Seawater Analysis.* Queen's Printer, Ottawa, Canada: pp. 279–282.

STRICKLAND, J. D. H., EPPLEY, R. W. and ROJAS DE MENDIOLA, BLANCA (1969). Phytoplankton populations, nutrients and photosynthesis in Peruvian coastal waters. *Bol. Inst. Mar. Peru*, **2**, 4–35.

VACCARO, R. F. (1969). The response of natural miocrobial populations in seawater to organic enrichment. *Limnol. Oceanog.*, **14**, 726–735.

VINOGRADOV, M. E. (1962a). Quantitative distribution of deep-sea plankton in the western Pacific and its relation to deep-water circulation. *Deep-Sea Res.*, **8**, 251–258.

VINOGRADOV, M. E. (1962b). Feeding of the deep-sea zooplankton. *Rapp. Cons. perm int. Explor. Mer.*, **153**, 114–120.

VINOGRADOVA, N. G. (1962). Vertical zonation in the distribution of deep-sea benthic fauna in the ocean. *Deep-Sea Res.*, **8**, 245–250.

WRIGHT, R. T. and HOBBIE, J. E. (1965). The uptake of organic solutes in lake water. *Limnol. Oceanog.*, **10**, 22–28.

ZOBELL, C. E. and HITTLE, L. L. (1969). *J. Oceanog. Soc. Japan*, **25**, 36–47.

On the interstitial fauna of the continental shelf of the state of Paraná (Preliminary Report)

HANS JAKOBI

Department of Zoology, University of Paraná
Curitiba, Paraná, Brazil

The interstitial sand biocoenosis is one of the most constant biotopes known. Research carried out during the last ten years has shown that its extension is quite large throughout the continental shelf of the state of Paraná. So we found it interesting to start a microfaunistic analysis of Paraná's meiofauna covering the continental shelf in accordance with the microfaunistic research of the International Biological Programme.

This substrate permits the development of an interstitial fauna due to its microspacial constitution. Herein we find a number of different species belonging to several classes of the hydrofauna which have been studied very little in South America. Recently we concluded a series of papers about ecological patterns of the fresh water interstitial fauna (Jakobi 1962, 1969; Jakobi and Marinoni 1965; Jacobi and Souza 1968). Based on those results we felt encouraged to follow the same methods and principles already applied in the analysis of limnic microfauna for a deeper study of the ecological situation of the meiofauna of Paraná, expecially on the continental shelf.

THE GEOCONSTANCY

The physical composition of the shelf substrate formed by sand gravel layers presents a special resistance combined with a determined granulometric value. All sediment samples were measured by the granulometric test standard method[1]. The relative frequency of granulometric fractions is fundamental for the build up and coenotic composition of meiofaunal populations. Samples collected from several check spots on the continental shelf proved that there are large areas with substrates within the values of 1–4 mm. These are considered excellent for the development of a faunal assemblage rich in species and individuals. Granulometric values higher than 0.2 mm are found mainly between the strip from Pontal do Sul up to the Sai River mouth. On the other hand diameters lower than 0.2 mm are clearly unfavorable for the development of large interstitial coenosis, especially when they are mixed with clay, mud, or other space obstructing material, as occurs mostly in areas in front of the "Ilha de Superagui", in its northern part.

These granulometric values also represent a biological limit for the normal crowding of the metazoan meiofauna. It is usually found associated with a high concentration of Protozoa populations, mainly Ciliata, as well as other small individuals in these minispaces. We also noticed that metazoan meiofauna florished on the entire continental shelf in sediment greater than 0.2 mm. In all these sand areas we found a preponderance of harpacicoid copepods and nematodes, constituting the most important members of the interstitial community. However, we also found here a series of associated species belonging to Polychaeta, Gastrotricha, Turbellaria, Nemertini and other less important representatives in this labyrinthical system of microcaves. Recently discovered were representatives of a new class, the Gnathostomulida (Ax, 1966); characteristic forms occuring in depth to about 20 meters. More data were obtained during a series of analysis I made in 1962 (Jakobi 1962) which showed that the fauna has many specialized morphological and physiological adaptations. These can be detected directly, are related to the kind of substrate, and may be tested for statistic significance (X^2 and t).

Considering the evidence of geoconstancy, specialists in oceanography have payed very little attention to this important link of the ocean life cycle which is necessary for the normal raising of macrofauna representatives such as shrimps, prawns, crayfishes, lobsters and fishes of commercial value. Research on the energetic role of meiofauna is yet in its infancy, in spite of existence of various papers on phytoplankton production intensity, which can be easily measured by chlorophyll concentration, or on the determination of P, N, oxygen or salinity related to the productivity of certain water units.

THE THERMOCONSTANCY

The interstitial temperatures found in Paraná's latitudes corresponds to the Tropic of Capricorn and it varies on the continental shelf by only a few degrees (21–25°C) during the year. In summer there is generally very little diurnal variation, especially deeper than 15 meters. These small thermal oscillations illustrate the relative thermoconstancy of the interstitial waters where the temperature cannot act as a factor of natural selection. The continuity of microfaunistic reproduction within the sand layers is of extreme importance for the natural renewal of shrimps and crayfish banks to a large extent along Paraná's coast.

THE CHEMICAL CONSTANCY

Like the other parameters mentioned above the salinity and pH determinations revealed only small oscillations for the majority of the sand areas studied. For depths greater than 10 m, the sand banks act as a sponge retaining large quantities of water which are independent of tidal water movements. Only the most exposed sand layers are washed and welled by local currents. Water of higher salinity and density is retained in the interstices during the time of low tide, even near the mouth of bays and rivers (Bay of Guaratuba and Bay of Paranaguá). This can be considered as another factor contributing effectively to the constancy of interstitial life and neutralises the occasional or seasonal influences of strong rainfalls. The rainfalls occasionally cause high salinity oscillations at the surface water and kill less protected biocoenosis like phytoplankton pools. On the other hand the interstitial environment continously absorbes the marine detritus brought from the rivers, estuaries, and/or that which is sedimented as dead plankton in calm off shore regions. Determinations of the quantities of biomass and organic substances showed that the water coming from Paraná's two bays (Paranaguá and Guaratuba) are very rich in organics constantly supplying the shelf meiofauna with nutrients. Meiofauna primarily eats small detritus particles coming from dead animals or plants. This fact can be evaluated as one more contribution to the biocoenotic constancy of these communities. In this way the populations metabolize substances already deteriorated and reintroduce

14*

them into the food cycle. Submarine sands with a rich microfauna are frequently visited by crayfishes and young fishes which are commercially important during at least a certain period of their growth.

THE BIOCONSTANCY

This characteristics is very important for understanding the unusual dynamic balance of these populations. As the substrate size predetermines its inhabitants, it is common that the representatives have a very slender and soft body with special adaptations for locomotion, digestion, respiration, orientation and reproduction. An unprotected body when exposed easily serves as food for the above mentioned young forms of the macrofauna and, considering that the qualitative composition of the meiofauna of a determined sand area is quite constant, the young fishes can always find the same type of food at the same place without great danger of extermination. Many interstital forms have special adaptative organs; for instance the haptic ones which fix the body on the gravel surface when any strange movement arises. Observations made in the laboratory clearly showed the tendency and capacity of harpacticoids (*Ectinosoma* sp.; *Stenocaris* sp.) to go immediately back to the sand surface when they were suddenly forced out by applying a localized water current. However, in general, that interval may be sufficient for some young fishes to catch some slow harpacticoids. In the open sea fishes hanging over the meiofaunal sand banks may continuously disturb the sand layer surface by means of the rhythmical movement of their fins. It is also evident that the special characteristics of the interstital fauna are responsible for its high resistence against phenomena like "upwelling". In all these years we have not found any typical representative of sand layer inhabiting fauna in upwelled surface water.

Acknowledgement

I am pleased to acknowledge our thanks to Prof. Dr. J. J. Bigarella, Geological Department, University of Paraná.

References

Ax, P. (1966). Eine neue Tierklasse aus dem Litoral des Meeres-Gnathostomulida. *Umschau i. Wiss. u. Technik* **1**, 17–23.
Jakobi, H. (1962). Harpacticoidea (Cop. Crust.) e Syncarida troglobiontes. *Zoologia* **21**, 1–92.

JAKOBI, H. (1969). O sigbificado ecológico da associação *Bathynella-Parastenocaris* (Crustacea). *Zoologia* III, 7, 187–191.

JAKOBI, H. and MARINONI, R. (1965). Ueber die Zuechtung v. *Attheyella* (*Chapuisiella*) *derelicta* Brian 1927 in granulometriertem Mesopsammal. *Arch. Hydrobiol.* 61, 1, 86–99.

JAKOBI, H. and SOUZA, E. A. (1968). Contribuição ao conhecimento da pesca na Paraná: *Zoologia* II, 14, 329–358.

Marine pollution
and its biological consequences

P. KORRINGA

Netherlands Institute for Fishery Investigations,
IJmuiden, Holland

Resumen

A seguinte definição do têrmo "poluição marinha" é agora aceita internacionalmente:

"Introdução pelo Homem, de substâncias no ambiente marinho, as quais têm efeito deletério sobre a vida marinha em geral, a saúde humana e sobre as atividades marinhas, incluindo a pesca, além de impedir o uso da água do mar e reduzir as amenidades".

Embora o volume ocupado pelos Oceanos seja tão grande de forma que as mudanças ecológicas devidas a descarga de "lixo" difìcilmente são notadas, a experiência prática mostra que podem ocorrer consequências sérias do ponto de vista biológico.

Isto se deve, principalmente, ao fato de que as águas de pouca profundidade e estuários próximos a costa, são de importância fundamental para a aquicultura em geral, sendo ainda onde se densenvolvem os estágios larvais de muitos peixes, cujos estoques podem sofrer diminuição com os efeitos da poluição. Existe uma grande variedade de substâncias poluintes, algumas prejudiciais à saúde humana, outras contaminando ostras, com efeitos desastrosos sôbre os organismos marinhos e seus predadores, incluindo o homem. Portanto, uma discriminação cuidadosa é um pré-requisito essencial para a planificação da descarga de material no Oceano.

An ever increasing amount of waste is produced by man, and especially in the more densely populated areas of the world adequate elimination of this waste is a very serious problem, indeed. Part of the liquid and semi-liquid

215

domestic waste discharged by sewers is bio-degradable. When discharged in bodies of fresh water biological processes lead to a gradual mineralization of the organic components of the domestic waste, and in due course the water is as clean and odorless as before, be it somewhat richer in nutrients. Not so very long ago one firmly believed that such self-purification processes would take care of elimination of any quantity of sewage discharged in lakes and streams. One had seen that the bio-degradation capacity of smaller and stagnant bodies of fresh water is limited, for the amount of oxygen dissolved in water is rather small and easily depleted during intensive breakdown of organic waste, which leads to switching over to anaerobic microbiological processes which work slower and often produce evil smelling gases. But soon it turned out that even lakes and rivers have only a limited capacity for selfpurification. Where in addition to domestic waste an ever increasing amount of industrial waste products, which often are of a non-bio-degradable nature, is discharged in the rivers those turn into evil smelling open sewers, devoid of fish, no longer suitable for recreational purposes such as swimming and sailing, and hardly for irrigation of arable land and for infiltration of the areas where drinking water is produced. The river Rhine is a notorious example.

The gradual deterioration of the water quality in lakes and streams is a reason for growing concern. Technically it is possible to apply special treatments to waste of various description in order to render it inoffensive and to eliminate poisonous constituents, so that the liquid finally discharged is hardly distinguishable from pure fresh water. But this requires considerable investments and appreciable running expenses. As long as governments do not compel cities and industries to apply such purification techniques one will look for cheaper ways to get rid of waste. When lakes and rivers have a too limited capacity to accept all the waste produced, would discharge into the sea not offer a reasonable solution of the problem? Bio-degradation processes will certainly see to a gradual mineralization of domestic waste in the sea as they do in fresh water. Since it is the volume of water which counts in such processes and not the surface area, it can safely be assumed that— since all the sea water combined offers such an enormous volume in comparison with all fresh water combined—the capacity for bio-degradation of domestic waste is almost unlimited in the sea. Further, it is usually assumed that industrial waste of non-degradable nature will, if properly dispersed in the sea, never reach lethal concentrations, not even for the most sensitive marine organisms. Those considerations have already led to construction of pipe-lines for discharge of waste into the sea, to an increasing tendency to build industrial plants at the sea shore for a cheap and easy disposal of waste, and to construction of special tankers to transport industrial waste to the sea.

It soon turned out, however, that discharge of waste into the sea does not necessarily provide such an easy and safe solution as one thought. A small and shallow sea as the North Sea contains 54,000 km³ of water so that dumping 54,000 tons of any material in it would, if properly dispersed, only lead to a concentration of $1/\mu g/liter$, a concentration so low that hardly any organism would suffer from it. Dispersion appeared to form the bottle neck in this area. Copper sulphate clandestinelly dumped at the low water line near Noordwijk, Netherlands, killed fish and invertebrates in a narrow strip of coastal water. The poisonous water of a salinity and temperature differing somewhat from that of the water further off-shore, and therefore not easily mixing with it, traveled with the tides northward along the coast and entered the Wadden Sea, threatening the huge quantities of mussels cultivated there. A similar event, originating from a quantity of the powerful insecticide telodrin traveling with the tides from Rotterdam to the Wadden Sea, led to the almost complete extermination of the once so numerous Sandwich terns there. Evidently one cannot count on a quick and thorough dispersion of dissolved waste in this section of the North Sea.

Floating waste, among which oil should be mentioned in the first place, is transported by wind, and is often washed ashore, polluting the beaches and thereby deteriorating the amenities. The "Torrey Canyon" disaster did not fail to focus the attention of the general public on the problem of marine pollution.

International organizations such as FAO, ICES, and IOC did not fail to realize that marine pollution is a problem of growing importance, and did what is within their power to search for a solution. IOC, the Intergovernmental Oceanographic Commission of UNESCO, set up a working group on marine pollution, to advise member-countries what further scientific research is required in this special field. One of the tasks of this working group was to produce a good definition of the term "Marine pollution". This runs "*Introduction by man of substances into the marine environment resulting in such deleterious effects as harm to living resources, hazards to human health, hindrance to marine activities including fishing, impairment of quality for use of sea water, and reduction of amenities*". ICES, working on a regional basis, produced recently detailed factual reports on marine pollution in the North Sea and in the Baltic Sea, and FAO is preparing a world congress on marine pollution for the year 1970.

It soon became clear that impact of waste discharge in the sea is greatest in inshore waters. This not only because dispersion is slow in the coastal water fringe with its differing temperature and salinity, but especially since shallow inshore areas are very often important spawning and nursery grounds for a variety of fish and shellfish. Further, farming of shellfish and seaweed can only be practiced in well-sheltered coastal water, and are not the amenities for marine recreation typical for the coastlines.

The solution seems rather simple: if one decides to discharge waste into the sea, this waste should be taken well offshore with the aid of pipelines or by tankers, the further away the more dangerous or contagious the waste is. By preference discharge should be carried out beyond the continental shelf, or, for more toxic material, packed in drums on the bottom of the ocean, far away from fisheries resources. In reality things are not so simple, for legally speaking this ground has not yet been ploughed. Beyond the territorial waters nobody can give rigid prescriptions as long as one has not adopted an international convention on marine pollution. No small wonder that the industry hesitates to go to great expenses for taking its liquid and solid waste far away, where it can do little or no harm to aquatic resources and to the amenities so direly needed for recreation. Since it is economically unsound to go to greater expenses than is really necessary to keep the marine environments clean, a certain discrimination will be necessary in a regimentation for discharge of waste into the sea, so that one can get rid of the more harmless substances at lower expense than more dangerous waste. Such regulation should be based on scientific information concerning the consequences of discharge in sea of waste of a given nature. A concise review of the various categories of pollutants can serve to give an impression of the biological consequences of their discharge into the marine environment:

1. *Domestic waste,* rich in organic material, requires a considerable amount of oxygen for its bio-degradation. Therefore the water will often be black and evil smelling in the immediate vicinity of the site of discharge, and soft black stinking mud accumulates there on the bottom. But even at a certain distance from the point of discharge consequences can be serious since domestic waste often contains contagious material. Bathing in diluted sewage can therefore easily lead to infestations of various description, brought about by pathogenic micro-organisms (e.g. typhoid fever), by

viruses (e.g. hepatitis, influenza, "colds"), and even yeasts (vagina irratation). The risk is greater where tides and currents are of little importance, as is for instance the case on the Mediterranean beaches. Seaside resorts mushroom, but usually do not invest money in purification plants for treatment of their own domestic sewage when not compelled to do so by governmental prescription. Therefore "short circuits" occur too often in places where people concentrate for marine recreation.

Indirect infestation can be traced to consumption of raw shellfish from water containing diluted sewage. Oysters and mussels filter large amounts of sea water in search of food and oxygen, and thus may concentrate contagious elements from sea water polluted by domestic sewage. In many countries, governmental prescriptions safeguard the public from eating contaminated shellfish. This involves loss of extensive areas of formerly productive shellfish beds. When pollution increases shellfish farming has to move further and further away from the polluted sites, but since shelter is an important factor in shellfish cultivation this often means ultimately that shellfish farming is no longer possible in the area under consideration. Where the water is only slightly polluted, shellfish can still be grown, but should pass a special shellfish purification plant before the required certificate of cleanliness will be delivered.

One often hears the slogan "farm the sea", but in reality indiscriminate discharge of domestic waste has already led to destruction of a noteworthy percentage of grounds suitable for shellfish farming, of potential aquatic resources.

It has often been said that discharge of domestic waste and similar non-poisonous bio-degradable material has also a positive aspect since bio-degradation leads to production of nutrients, among which phosphates and nitrates, often the limiting factors in plankton development. It can and does happen that enrichment originating from discharge of domestic waste locally leads to a richer plankton and eventually to more fish and shellfish in the area under consideration; but it happens more often that such discharges lead to eutrophication, to an excess of nutrients, which in their turn may lead to waterblooms of a variety of phytoplankton organisms among which dangerous dinoflagellates such as *Gonyaulax*, ensuing in "red tide" and paralytic shellfish poisoning. But also lighter degrees of eutrophication can have a deleterious effect such as lack of settlement of oyster spat and poor growth and fattening of shellfish in waters too rich in phytoplankton. For a body of water such as the North Sea one has amassed sufficient evidence to demonstrate that food is not the limiting factor in the production of fish and shellfish. Experience gained during two world-wars has shown conclusively that much greater stocks of fish can live there without showing a sign of reduced growth through shortage of food. Enrichment of the North Sea area therefore is no valid excuse for discharge of waste.

2. *Industrial waste* contains material of a widely different nature. Much of it is not bio-degradable and therefore it is not realist to try to lump all this waste, as is done for domestic waste, under the heading "Biological Oxygen Demand" or "Inhabitant Equivalents". Mineral acids, among which sulphuric acid, are among the bulk products of industrial waste to be disposed of in the sea, since fresh water has a too limited capacity for safe absorption of acids. Sea water has such a considerable buffering capacity that, if efficiently spread over an offshore area of reasonable dimensions, those acids will hardly affect the pH of the sea water. Specially equipped tankers are used to this effect. The elements dissolved in the waste acids often are reason for greater concern than the acids themselves. Organic acids, often possessing a strong odor, should on the other hand not be discharged in noteworthy quantities in shallow seas, for fish and shellfish may become tainted by them, which may upset the whole fish market.

Heavy metals, sometimes dissolved in waste acids, should be treated with considerable caution. Though the concentration to be expected after discharge in the sea may be well under the lethal level, serious aftereffects are by no means imaginary. A variety of invertebrates, among which bivalves, sponges, and ascidians, are filter feeders, and such filter feeders often accumulate certain elements to an amazing degree. That oysters contain per gram more than 100,000 times the amount of copper found in one gram of sea water, is nothing unusual. For zinc and mercury figures can be higher still. Ascidians, on the other hand, are renowned for their accumulation of vanadium. It always concerns elements which are of rare occurrence in natural sea water and which are taken up from the sea water by some biochemical process. In bivalves it is the biochemical collection of calcium ions needed for shell construction which inevitably leads to ingestion of other positive double valenced ions, which cannot always easily be excreted and are then stored in inoffensive form in the connective tissues. Evidently the bivalve itself does not suffer from those stored surpluses, but if a predator ingests molluscs which are unusually rich in copper or mercury death may follow, for its digestive processes bring the metals back in ionic form.

The Minamata disease in Japan, which killed and disabled a great number of people, is such a case of mercury poisoning brought about by shellfish which has accumulated considerable quantities of mercury.

Sea water has been remarkably constant in chemical composition throughout the aeons and therefore many of its denizens are so vulnerable against changes, against which they never had to defend themselves. One of the consequences could be called "biochemical inflation", described above, that always concerns elements rare in natural sea water and playing some role in the physiology of marine organisms. Very similar things may happen when pesticides of certain description enter the marine environment, either in the drain-off from the land or by purposely dumping in sea of superflous batches. Some types of pesticides, generally indicated as "hard" insecticides, are persistent. Chlorinated hydro-carbons such as DDT are notorious among the hard insecticides. They resist all biological and chemical attacks under natural conditions and therefore one must count on their perpetual presence in the natural environment. In the sea these fat-soluble products can sooner or later be located in the fatglobules of diatoms and in due course in the fat of copepods. As such they do little harm for the time being, but here again disaster may strike when predators consume a fair number of prey-organisms containing pesticide-loaded fat and make the dangerous product free by digesting the ingested prey. The further in the food chain, the higher the level of the pesticide will be through the process of biological accumulation. The same phenomenon is observed on the land where at the end of the food chain birds of prey fall victim to persistent pesticides. For marine organisms man may stand at the end of the food chain.

It would be a very wise thing if the use of persistent pesticides could be forbidden in all countries of the world. For the time being we still have to count on their presence in the natural environment in ever increasing quantities. Dumping in the sea of such products should never be allowed. It would be best to destroy unsalable batches by incineration. Special boats are already in use to incinerate such products at sea.

Radioactive waste in very diluted form is locally discharged into the marine environment. Scientific research is carried out to follow the where-abouts of potentially dangerous material. In this, radio-active elements with a rather long "half-life" are of first importance. Of those, the elements which occur only sporadically in sea water and which do play some role or other in biochemical processes require special attention. Radio-active isotopes are biochemically of the same value as the non-radio-active elements. Therefore biological accumulation is only feasible in elements which occur only sporadically in natural sea water so that discharge of a noteworthy quantity of the radio-active isotope can lead to its biological accumulation. Fortunately the waste products of nuclear power plants are largely radio-active elements such as niobium and zironium, which are biochemically accumulated by marine organisms.

In Europe special research on the pathways of radio-active matter in the marine environment, included their possible accumulation in the food chain, is carried out in a specialized laboratory at Fiascherino, Italy, under the guidance of CNEN and EURATOM. The results obtained will be used in future regulation of discharge of radio-active waste into the sea. This is a good example of scientific research which preceeds large-scale discharge of waste. If any element is indicated as potentially dangerous, efforts can be made to eliminate that element from the liquid waste before the latter is discharged into the sea.

Petro-chemical products, among which is crude oil, are for the general public the most notorious pollutants of the sea. Despite international agreements to keep the sea as clean from floating oil as possible many beaches suffer badly from oil pollution, and year after year countless sea birds die a cruel death through contact with floating oil fields. As long as the oil is floating the denizens of the sea have little to fear. Petro-chemical products are as a rule not really poisonous. The "Torrey Canyon" disaster has demonstrated that under water more victims have been made by the detergents, so liberally spread on the oil-polluted beaches, than by the oil itself. Tainting of fish and shelfish is, however, to be feared in oil-polluted water, and may upset the fish market in no little way.

Hot water discharged by power plants can be considered among the few waste products which may have a positive effect in sea. Acclimation of valuable species requiring higher temperatures in some phase of their life cycle has already been achieved in this way:

Mercenaria mercenaria, the American hard shelled clam, thrives now in the Southampton region, since its reproduction has been made possible by the higher water temperatures. Cultivation of valuable species of fish and shellfish can eventually also profit by the local warming up of sea water.

Summarizing it can be concluded that discharge of waste into the marine environment often leads to a conflicting situation. For reasons of economy a discharge closely inshore, by preference in estuaries, bays or lagoons, is preferred. On behalf of both fishery and recreation discharge well offshore, whenever possible beyond the continental shelf, is, however, highly desirable. Technically this is possible, but expenses will be very high. Contrary to the opinion of the general public it can safely be assumed that big industry has no objections to purification of liquid waste or its transportation to sites where safe disposal is possible, provided all countries adopt one and the same regulation for waste disposal, so that false competition is eliminated. It is the consuming public which will ultimately pay for the inevitable extra expenses. In order to avoid unnecessary expenses all potential pollutants should be carefully classified on a scientific basis and according to the character of the consequences ensuing from their discharge into the sea. They should be placed in categories for each of which a certain type of disposal will have to be prescribed. The prescriptions may range from: inshore discharge (e.g. warm water) via offshore discharge (e.g. domestic sewage, mineral acids) to discharge beyond the continental shelf (heavy metals, organic acids). More dangerous materials should be deposited on the bottom of the ocean, packed in drums (e.g. some radio-active waste), whereas the most dangerous and persistent material (e.g. persistent pesticides) should never be brought to the sea, but should be disposed of in another way. It is clear that only through an international agreement on the regulation of discharge of waste into the sea one can hope to solve the serious problems ensuing from indiscriminate waste disposal in the marine environment. An international convention on marine pollution should be framed, if possible on mondial basis, when necessary on a regional basis first. Material for framing such a convention is now being adduced by a Joint Group of Experts on the Scientific Aspects of Marine Pollution (GESAMP) and there is every hope that an international convention, comparable with the convention on oil pollution of the sea by IMCO, will once be adopted and put in practice under a rigid system of international control. Thus far the separate governments are reluctant to impose extra expenses on their national industry, for they fear that such industries might move to other countries with a less stringent regulation. A good international regulation can safeguard the industry and in the same time prevent both destruction of a variety of aquatic resources and deterioration of the amenities.

Distribution of benthos on the south west coast of India

C. V. KURIAN

Department of Oceanography
University of Kerala
Cochin-16

Resumen

Estudos do bentos da costa sudoeste da Índia estão baseados em coletas com dragas e "pegadores" de fundo, em 150 estações distribuidas na plataforma continental que se estende de Mangalore ao Cabo Comorin, cobrinda uma área de 30.000 km². Juntamente com as amostras de fundo foram realizadas coletas para hidrografia, na superfície e próximas ao fundo.

Os resultados das pesquisas mostrátam que existe uma relação bem marcante entre a natureza do depósito e a densidade do bentos. Areia fina com pequena percentagem de silte constitui o melhor substrato para a macrofaune, constituida principalmente por poliquetas e crustáceos.

Existe uma tendência geral na diminuição da densidade do bentos com a profundidade, mas a região próxima à plataforma continental é geralmente mais fertil que a zona costeira. A densidade da fauna é máxima nas proximidades das praias, nas entradas de estuários e costas rochosas, onde a biomassa, constituida por bivalvos, atinge 400 g/m². O pêso úmido total, incluindo conchas, pode atingir até 10 kg/m². Números semelhantes ocorrem próximo a linha de 182 m.

A biomassa bêntica nas regiões de lodo é alta, atingindo também, 400 g/m². Uma comparação das amostras de peixes de fundo e camarões, mostra que o máximo ocorre na regiões de alta biomassa bêntica.

INTRODUCTION

One of the prerequisites for the development of demersal fisheries is a thorough knowledge of the conditions of the sea bottom. The lack of sufficient data on the nature of the sea bottom has been a great handicap for the development of the bottom fisheries of the Indian region. Besides the nature of substratum, the hydrographical features of the area also plays an important part in the distribution of benthos.

Though observations on species or groups of marine bottom animals have been made as early as the later part of the eighteenth century, quantitative work using suitable sampling equipment is comparatively recent, begun only with the investigation in the Danish waters by Petersen (1911) who based his evelution of the demersal fish resources of the sea on estimates of benthos biomass. The standing crop of marine food resources is not only important to the demersal fishes which directly feed on them, but also to many pelagic species that restrict to shallow waters during some period of their life. The composition of the benthic community and the standing crop of benthic animals are determined by the conditions of life on the sea bottom. Thus, if it is possible to establish the relations between the benthos and the environmental factors it is easier to assess the standing crop by measurement of these factors. Petersen and Johnson (1911), Blegvad (1917), Sparck (1935), Thorson (1957) and others classified benthos into 'infauna' and 'epifauna' and also into 'communities'. But Stephen (1923) preferred the concept of zonation and he refers to 'shore zone', 'coastal zone' and 'off shore zone' depending upon the location of the place of collection. Jones (1950) is of the opinion that the bottom living fishes and crustaceans should also be regarded as part of the bottom communities.

The early investigations of the sea floor by dredging is summarised by Holme (1964). Most of the early studies were directed towards the discovery of species and enumerating them, but during the beginning of the present century interest in the benthos became more ecological with particular reference to benthos as a source of fish food. Quantitative studies were initiated by Petersen and Johnson (1911) who used a grab of 0.1 m² area for bringing up the sample. This instrument though modified later by Smith and McIntyre (1954) remains the basic standard tool for quantitative collection of benthos from soft grounds. As the weight of the instrument alone helps in the penetration to the bottom the instrument gives satisfactory results only in soft mud and muddy sand. However, Smith and

McIntyre grab widely used in the British waters and also now used in the Danish seas is found to give better results owing to the spring loaded mechanism which helps in driving the jaws of the grab into the bottom. But even now for the sandy and gravelly bottom only dredges are found to work satisfactorily for collecting the fauna.

Though many factors affect the distribution of benthos in the sea, the significant environmental factors that may cause the fauna in the littoral region to become segregated into groups are temperature, salinity and the nature of the bottom deposit. These three factors interact and when two are fairly uniform over a large area, the third will be relatively more important. Thus when salinity and temperature are uniform, bottom deposit will be the controlling factor. In very deep water all the above three factors may be uniform over a very large tract and the fauna may vary very little. In shallow coastal waters and in enclosed situations with a large influx of fresh water, salinity will assume great importance.

Sparck (1935) has shown that the average number and weight of living organisms has a correlation between production, climatic factors and also with fish population. It is likely that the bottom animals per unit area may differ greatly on grounds which are separated by only a short distance.

The earlier workers (Mortensen 1925, Remane 1933) divided benthos into 'microbenthos' and 'macrobenthos' considering the size. Mare (1942) coined the term 'meiobenthos' to distinguish the organisms of intermediate size. The microbenthos include bacteria, diatoms, and most protozoans and are usually separated by bacteriological techniques. The macrobenthos constitute organisms which are separated by sieves of 2 mm to 0.5 mm mesh (according to different authors). It is desirable that the lower limit for the separation of macrobenthos is fixed as 0.5 mm, so as to be uniform with all workers in the field. Again, the meiobenthos forms organisms that pass through 0.5 or 1.0 mm sieve and having a lower limit of 0.04 or 0.1 mm (McIntyre 1969). Though the above divisions are only arbitrary, it is desirable that the upper limit for meiobenthos is fixed as those passing through 0.5 mm sieve and lower limit as those which are retained in 0.04 mm sieve.

A general interest on the small groups of animals came only with the works of Remane (1933) who had his investigations in the Kiel Bay. As the meiofauna constitute mainly Nematoda, Harpacticoida, Ostracoda and a group of very small organisms including larval forms of macrofauna, it is essential that the collections are made with great care. As the upper few centimeters of the deposit contains most of the meiofauna it is essential to collect undisturbed samples using grabs, corers and other special instruments like Muus trap (Muus 1964). Rees (1940) found the maximum density of meiofauna in the upper part of the intertidal zone and in the sub-tidal as well as beyond the shelf the meiofauna is rare. In most cases the meiofauna is about 30 to 190 times more numerous than the macrofauna (McIntyre 1969). In the intertidal mud flats the meiofauna even exceeds a thousand times. A large proportion of meiofauna individuals appears to be free from predation by animals of higher trophic levels as they could easily bury into interstitial spaces in the sand and mud.

MATERIAL AND METHODS

The present study is based on a series of benthos and bottom deposit collections taken from the continental shelf of the S.W. coast of India extending from Mangalore (12° 50′) to Cape Comorin (8° 00′) during 1943–1947 and 1958–68. Data from 150 stations along 17 profiles and intermitant stations are made use of. Quantitative collections of benthos and deposits were obtained using Petersen and Van-veen grabs from the Research Vessel *Conch*. A naturalist's dredge and a 6 ft. beam trawl were also occasionally used for collecting additional data about the fauna. The grabs were found to work satisfactorily in a varied type of deposit ranging from soft mud to muddy sand. However in sandy and gravelly deposits as the grabs brought only very little sample, bag net and dredges were employed for collecting deposit and fauna in addition to the grabs.

A sample of the deposit was preserved for later detailed analysis and the macro-fauna was separated using 0.5 mm mesh sieve. The wet weight of the fauna (formalin preserved) after removing the shells in case of molluscs and tubes of polychaetes was determined in each case. The average of 2 grab samples for each station was taken into account for calculating the biomass. The bottom deposits collected were later analysed mechanically using standard sieves and the different grades were expressed as per Wentworth's size classification (Krumbein and Pettijohn 1938). Nansen bottles were used for taking hydrographical data from standard depths and also from very near the bottom and the salinity, temperature and disolved oxygen were determined as per standard procedure. Van Veen and core samples were used for meiobenthos study. Some hydrographic data collected by M. O. Kristensen and R. V. Kalva (Jones 1962) were also made use of.

HYDROGRAPHY

Normally the surface temperature in the shelf beyond 20 m depth varies from 23 to 31°C, maximum being observed during the summer months March–May and very rarely reach 32°C. Beyond the shelf the variation is less, the range being 26–31°C. The variation in surface salinity in the shelf byond 20 m is not considerable (30.5–35.5$^o/_{oo}$). Beyond the shelf it varies from 33.7–35.84$^o/_{oo}$. But within the 20 m depth the changes in salinity are more prominent owing to the influx of fresh water from rivers and canals. Throughout the coast a number of rivers and estuaries open into the sea (Fig. 2) and the monsoon rains have great influence in lowering the salinity in the inshore waters. In the 3 to 20 m zone the surface temperature varies from 25–30.1°C, the maximum being in April and the minimum during August–September. The salinity near the shore (4 m) may be as low as 6$^o/_{oo}$ in July, whereas the highest (34.9$^o/_{oo}$) occurs in April. The differences near the 8 m line are 10.72 to 34.6, at 15 m 13.12 to 34.6 and near 20 m 17.5 to 34.5$^o/_{oo}$.

The surface temperature and salinity do not generally suggest the nature of the bottom water which has direct influence on benthos. The temperature near the bottom is lowest during August–September and highest during April. It varies from 22 to 30°C within the 2 m line. Though there is considerable difference in the surface salinity in the coastal waters, the difference in bottom salinity is not so significant. Even at 3 m depth during July when the salinity at the surface is 6$^o/_{oo}$ the bottom salinity is 20.5$^o/_{oo}$. The highest bottom salinity in the 3 m to 20 m zone as occurred in April 1966–68 is more or less uniform and varies only from 34.5 to 34.8$^o/_{oo}$. But the lowest salinity which occurs in July varies from 20–31$^o/_{oo}$ as we proceed from 3 m to 20 m. This again shows that the influence of fresh water during the monsoon period reaches upto 20 m depth, which is about 4–6 miles (6.4–9.6 km) away from shore in the northern half and 2–3 miles (3.2–4.8 km) away towards the south.

Beyond 20 m depth up to 50 m normally the temperature varies from 29 to 27°C. But in some places temperature as low as 20°C has also been observed. However there is not much difference in salinity which normally varies from 34–35.5$^o/_{oo}$ and only rarely at some stations it reaches as low as 32$^o/_{oo}$. This shows that the influence of fresh water from rivers is negligable beyond 20 m depth. At 100 m line the temperature variation is from 18°C to 27°C and salinity change is less (34.3–36$^o/_{oo}$). At 180 m depth

(edge of the continental shelf) the temperature varies from 18 to 19.10°C
and salinity 35.1 to 35.84⁰/₀₀.

In the stations beyond 30 m depth there is not much change in the bottom
temperature from place to place during the same season as the influence of
the rivers and estuaries seldom affect the bottom temperature and salinity
in this zone.

According to Banse (1968) the coastal surface currents off the south west
coast of India set toward the south from February until late October or
November and are reversed during the remainder of the year. The annual
salinity minimum occurs in January before the reversal of the coastal
current. The temperature maximum occurs before the onset of the south
west monsoon in water of high salinity coming from north.

Upwelling off the south west coast of India starts with the onset of the
south west monsoon and it lasts through out the south west monsoon period.
Upwelled water off the south west coast of India can be very low in oxygen.
Concentration below 0.25 ml/l are common in the shelf (Banse 1959).
The depletion of oxygen in shallow depth even upto 10 m affects the distribution of bottom fauna. In September 1967 and 1968 similar phenomenon
has been observed at 10 and 15 m depth off Malipuram near Cochin. The
high rate of production in the shelf is during the south west monsoon when
the nutrient content of the water is high. Again it is observed that the
withdrawal of the cool water from the shelf during November–December
is a transition to low rates of production.

BOTTOM DEPOSITS

The nature of the bottom deposit plays an important part in the distribution
of benthos. So a detailed study of the deposits of the shelf region was made
along with the fauna.

The width of the continental shelf varies at the different profiles. Whereas
it is only 25 naut. miles at Anjenjo, it is 50–60 naut. miles off Cape Comorin.
The echographs of the bottom contour at right angles to the coast show that
the slope is gradual and drops to 100 fathoms (ca. 182 m) off Trivandrum,
Vizhingom, Quilon, Ponnani, Alleppey and Puvar within 21 to 27 miles
from the coast. But at Cochin, Calicut, Cannanore, and Mangalore the
sudden increase in depth is only after 35 miles (56 km).

Near the coast the deposit is generally sandy, the nature of the sand
differing according to locality, except in places where there is rock as at

Cape Comorin, Vizhingom, Quilon, Beypore, Tellicherry and Cannanore or mud as in the mud bank regions. This sand extends from the intertidal regions to 4–5 m depth or to still deeper regions according to locality. In some places south of Ayiromthengu and extending as far as Chavara, fine ilmenite sand is mixed with coarse and fine sand. Similar deposit occurs also at particular places near Cape Comorin. The rock near the shore, generally laterite or gneiss also extends from the shore to 3–5 m depths. They are generally found in patches.

The soft ooze-like mud and compact clay is a characteristic feature of the mud bank regions. These occur in patches in different places extending from north of Calicut (11° 30′ N, 75° 38′ E) to south of Alleppey (9° 15′ N, 76° 24′ E)

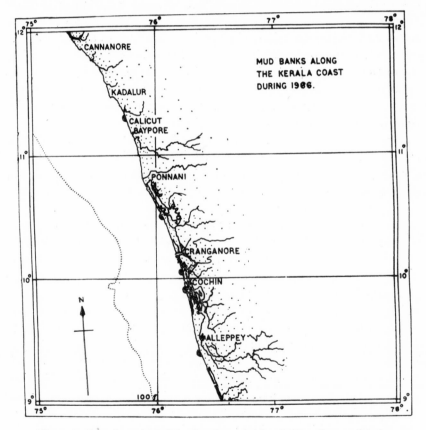

Figure 1 Location of Mud banks along the South West Coast of India (Kerala) during 1966

(Fig. 1). The mud banks extend from the low water level to 6–12 m depth. These banks are found in 3 groups, one near and around Calicut, the second near Cochin and the third near Alleppey-Purakkad region. Each bank may be 4 to 6 km in length along the coast and almost the same extense into the sea. Again, the mud banks may be grouped into temporary and permanent ones, the permanent ones appearing at definite places every year during the south west monsoon, whereas the places of appearance of the temporary ones may be different every year. Some of the banks after their appearance may extend their area or completely shift from one place to another. The mud in the banks may be silt or clay and sometimes as thick as 5 m. However these banks are known to navigators as places of safe anchorage for ships and to fishermen as rich fishing grounds, especially for prawns during the monsoons. There are many theories regarding the formation of the mud banks and the main works dealing with the mud banks are by Du Cane *et al.* (1938) Bristow (1938) Damodaran and Hridayanathan (1966). The reasons for the formation of the mud banks have not been clearly understood. However it may be said that during the south west monsoon, upwelling takes place along the south west coast of India, and the bottom mud which sometimes extends upto 15–20 m gets stirred up, the mud being lifted upto even very near the surface. By the interaction of a number of forces like currents, waves, river discharges etc. the mud settles down near the coast and forms the banks. The mud is of terrigenous origin derived from laterite rock, brought down by rivers and the fact that most of the banks occur very near the mouth of rivers or estuaries shows that the mud and alluvium brought down by the rivers pass through estuaries and get precipitated when it comes in contact with sea water thereby allowing it to settle in the inshore regions. The mud of the banks is of very fine grained type, most of which is of less than 0.002 mm diameter and the mud of some regions especially that of Alleppey is oily. The mud bank is of particular importance as it gives complete calmness to the water even in the most rough monsoon weather and the surf can be seen breaking at the boundary of the banks.

The mud banks play an important part in the biology and chemistry of the sea water. They are store houses of inorganic nutrients and organic matter (>4%) and this accounts for the high fertility of the region and production of phytoplankton and benthos and also the consequent increase in the pelagic and bottom fisheries. The works of Seshappa (1953), Seshappa and Jayaraman (1956) and Damodaran (unpublished) show that the mud banks act as reservoirs and liberate into the overlying water inorganic phosphate. Just before the formation of the mud banks the inorganic phosphate content of the water very near the bottom show values as high as 8 µg at/l (Damodaran and Hridayanathan 1966).

Beyond the 5 m line within Mangalore to north of Quilon extending upto 20–25 m the deposit is mainly silt and clay (upto 98 %). Fine sand appears in small percentage in the deeper stations. Generally the mud is grey in colour. But at 20 m off Mangalore and Cannanore soft greenish mud occurs. Off Quilon patches of clay and stones are present at 10 m. South of Quilon extending upto Cape Comorin sandy deposit predominates upto 20 m with occasional patches of rock at Anjengo, Muttom and Cape Comorin.

Beyond 20 m depth extending upto 30–50 m fine medium sand with mud occurs from Mangalore to Anjengo. But off Cochin the deposit formed of large percentage of silt may extend upto 40–45 m. But from Trivandrum to Muttom the mud fraction is less and the sand is more coarse. Off Cape Comorin, mud with small percentage of fine sand occurs in patches in between sandy grounds. This belt which extends from 8 to 19 km off shore is the best ground for prawns and bottom fishes (maximum within 20–30 m), prawns being most abundant north of Quilon.

Beyond 30–50 m depth extending upto 80–100 m the deposit varies at different places. The silt fraction is less, and stones occur off Mangalore, Cannanore, Calicut, Cochin, Quilon, Anjengo, Trivandrum, Vizhingom and Puvar, Greenish fine/coarse sand with silt occurs through out the region.

Beyond 80–100 m extending upto the edge of the continental shelf (180 m) the deposit is mostly fine grey/greenish sand with small percentage of silt and at some places mixed with coarse sand. Rock also occurs in patches as off Kasargod, Cannanore, Cranganore, Alleppey, Anjengo, and Vizhingom. Beyond the edge of the continental shelf the bottom is hard with greenish or greyish sand and very little silt and shell fragments.

BENTHOS

Near the barmouths where there is an aggregation of bivalves and poly-chaetes the benthic biomass shows a very high figure. At Cochin it is as high as 688 g/m² whereas at Ayiromthengu and Cranganore it is 290 and 350 g. respectively. The live weight of bivalves including shells come to about 10 kg/m². The common bivalves are *Meretrix spp.* and *Modiolus spp.* and the polychaete bed is formed by *Diapatra neapolitana* Delle Chiaje. Almost throughout the coast in the intertidal and subtidal region the biomass of

Fertility of the Sea

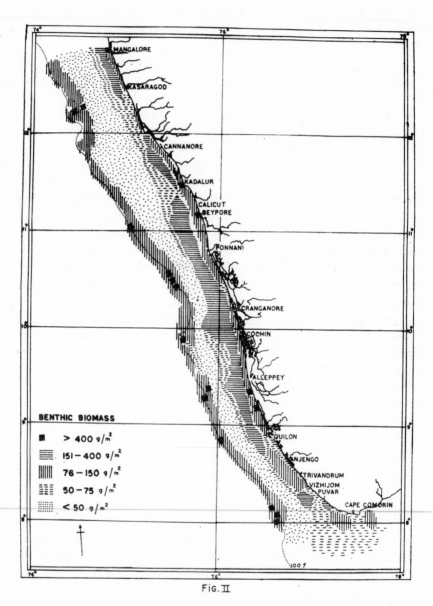

FIG. II

Figure 2 Distribution of benthos on the South West coast of India

bivalves is high. *Mytilus viridis* (Linné) forms extensive encrustations on the rocky shores.

In the mud banks the benthic biomass is high, especially after its formation. In the banks south of Alleppey a biomass of 400 g/m² has been observed during June–July, mostly constituted by polychaetes, *Turritella attenuata* Reeve, *Trypauchen vagina* (Bloch and Schn.) *Cynoglossus spp.*, prawns, Amphipods, and Ostracods. In this region the plankton also shows a high figure. But by September–October the macrofauna diminishes to <50 g/m². This region is the best known for the fishery of prawns and sole fishes.

In the Calicut mudbank region the biomass is comparatively less and may be as low as 80–100 g, but is richer than the bank off Malipuram (north of Cochin) which shows 30–50 g except when *Trypauchen* and *Turritella* specimens are caught. In the sandy coast at Malipuram when the mud bank appears, very near the low water level the tube dwelling polychaete *Sabellaria cementarium* Moore gives a figure >400 g/m² (wet weight excluding tubes). But the polychaetes vanish when the sea becomes rough after the disappearance of the mud banks.

Beyond 5 m depth in many places north of Quilon it is clay upto 15–20 m depth, sometimes extending upto 25 m. Here the macrofauna is scarce, though meiofauna is intense. In this region there may be variation in the fauna according to seasons and also localities depending on the influence of run off from rivers. The clay harbours few macro-fauna, the dominent groups being Molluscs, Ophiuroids, Nematodes and Crustaceans. The biomass varies considerably from 11 to 120 g/m², usually the higher figure occurring near the deeper regions where there is more sand. However, south of Quilon, the deposit is more sandy and the fauna is different. *Branchiostoma sp.* Cumacea, *Squilla* and crabs appear in small numbers.

Beyond 15–20 m depth extending upto 30–35 m and in some places like Cochin and Quilon upto 50 m, the benthic biomass is high. The deposit here is silty sand. The fauna consists of Polychaetes, Ostracods, Copepods, Mysids, Amphipods, prawns and Molluscs. This is the best ground for bottom trawling and prawn fishing, besides the mud bank regions.

Beyond 30–50 m extending upto 80–100 m depth the deposit is generally unsuitable for benthos and biomass is <50 g/m² sometimes as low as 0.5 g/m² in sandy deposit. However the average for this comparatively infertile zone is 8.5 g/m², higher than what is observed by Longhurst (1957) in Sierra Leone.

Beyond 80–100 m depth extending upto 180 m the deposit is more favourable, the changes in salinity and temperature round the year are negligible and the benthic biomass is greater. Perhaps the grazing by carnivores in this region is also less resulting in the increase of standing crop. The nature of the fauna varies according to deposit and the biomass varies from 15 to 150 g/m².

Beyond the edge of the continental shelf the biomass is more upto about 300 m in certain places and in the localities where 'ascidian mattings' occur the live weight reaches as high as 6 kg/m².

A perusal of the fishing charts given by Rao (1969) shows that the regions of density of ground fish and prawns more or less agree with the intensity of benthic biomass.

Damel and Mulicki (1954) have observed a definite relationship between benthos biomass (live weight) and commercial catch of fish. They have found that the high catches of 80, 28.8 and 24.5 kg/ha occur in Azov, Japan and North Sea where the benthic biomass show 313, 175 and 346 g/m² and the low catches of 1.5 and 12 kg/ha occur in the Mediteranean and White Sea where the biomass show 10 and 20 g/m² respectively. According to Filatova (1938) the edges of the Barents Sea and continental shelf show a biomass of 100–500 g/m². In regions where *Mytilus, Modiolus* and other shelled animals occur the live weight is still higher (16.5 kg/m², Sparck 1929).

In the south west coast of India, the major bottom fishery comes from the inshore regions within 20 to 30 m depth. This region shows a high benthic biomass which forms food of bottom fishes and prawns. The only other region which is more fertile in the inshore waters is the mud bank area. During the south west monsoon as already suggested upwelling occurs in the west coast of India and the water of low oxygen content appears very near the coast. The de-oxygenation of the near bottom water do not seem to affect much of the bottom fauna which are able to survive in water of low oxygen content. But the demersal fishes are found to shift from this zone. The harbouring of prawns and flat fishes in the mud bank regions may be due to the shifting of these animals from their natural habitat to favourable conditions. The mud bank regions are exceptionally calm during the monsoon and the liberation of phosphates and nitrates from the bottom mud by the churning of the mud helps in the better production of benthic organisms.

Banse (1967) observes that there was no relation between temperature of the water and total catch for Anton Brunn data from the Arabian Sea and that in terms of depth, fishing would be unprofitable between July and

November–December at depths greater than 50 m in the early part of the period and greater than 40 m in the later part. Where the distribution of fishes is affected by oxygen regime in the bottom water, the temperature of the bottom water is of little use in predicting catch. The lethal oxygen concentration of different species differ. It may be between 1.5 and 3.8 ml/1 in case of *Penaeus indicus*. But the deepwater prawn *Panaeopsis philippi sp.* Bate and spiny lobster *Puerulus sewelli* Ramadan tolerate even oxygen concentration lower than 1 ml/l.

Silas (1969) observes that the average catches of demersal fish by R. V. Varuna are 62.42, 256.87 and 273.65 kg/hr of trawling (ca. 9 hectres) for depths 75–100, 101–179 and 180–450 m respectively. This confirms that the zone 75–100 m depth with a sandy/rocky bottom and low benthic biomass is poorer for fishery compared to the zone 101–179 m, where the deposit is favourable and the benthic biomass is greater. However data on the bio-mass of bottom fauna and fish catch from the whole shelf region during all the seasons are required for arriving at definite conclusions.

Analysis of the data on the plankton hauls from the shelf region shows that it varies from 0.3 to 1.14 cc/C^3 (Indian Ocean Standard net) and that the intensity is more near the shore. Beyond the shelf it is <0.1 cc/C^3 (Rao 1969). The inshore waters between Calicut and Quilon show the maximum biomass of standing crop of plankton especially after the onset of the South West monsoon. Towards the south, the plankton is comparati-vely less. The abundance of plankton does not show any direct relation with benthic biomass, except that the inshore regions are more fertile for both plankton and benthos.

SUMMARY

Studies on the benthos of the south west coast of India are based on grab and dredge collections of benthos and bottom deposits from 150 stations distributed in the continental shelf extending from Mangalore to Cape Comorin covering an area of 30,000 sq. km. Along with the bottom samples, hydrographical collections from the surface and near bottom were also taken.

The results of the investigations show that there is a definite relation between the nature of the deposit and the intensity of benthos and that fine sand with small percentage of silt forms the best ground for macrofauna constituted mostly by polychaetes and crustacea.

There is a general trend in the decline of benthos with depth, but the region near the edge of the continental shelf is generally more fertile than the previous zone. The intensity of fauna is the maximum near the shore at the barmouths and rocky coasts where the biomass constituted by bivalves reaches >400 g/m^2. The total wet weight including shells may reach as high at 10 kg/m^2. Similar high figure is also found near the 100 f. line (ca. 182 m) at certain places where "ascidea mattings" are observed.

The benthic biomass in the mud bank regions is also high immediately after its formation, often reaching 400 gm/m^2. A comparison of the catches of bottom fish and prawns shows that the maximum catch comes from the region of high benthic biomass.

The aggregation of bottom animals in the mud bank regions during the south west monsoon period is suggested as due to migration from rough weather to a calm place of high production.

References

BANSE, K. (1968). Hydrography of the Arabian Sea shelf of India and Pakistan effects on demershal fishes. *Deep Sea Research* **15**, 45–79.

BANSE, K. (1959) On upwelling and bottom trawling of the South West Coast of India. *J. Mar. Biol. Ass. India I*, 1, 33–49.

BLEGVAD, H. (1917). On the food of fish in Danish waters within Skaw. *Rep. Danish Biol. Sta.* **24**, 19–71.

BRISTOW, R. C. (1938). *History of Mud banks I & II*, Cochin Govt. Press.

DAMODARAN and HRIDAYANATHAN, C. (1966). Studies on the mud banks of the Kerala Coast. *Bull. Dept. Mar. Biol. Oceanogr.* **2**, 61–68.

DEMEL, K. and MULICKI, Z. (1954). Quantitative Investigations on the biological bottom productivity of the South Baltic. *Rep. Sea Fish Inst. Gdynia* **7**, 75–125.

FILATOVA, Z. A. (1938). The quantitative evaluation of bottom fauna of the South Western part of Barents Sea. *Trans. Knipovich Polyar Sci. Inst.* **2**, 3–58.

HOLME, N. A. (1964). Methods of sampling the Benthos. *Adv. Mar. Biol.* **2**, 1964, 171–260.

JONES, N. S. (1950). Marine bottom communities. *Biol. Rev.* **25**, 283–313.

JONES, S. (1962). Oceanographic station list. *Ind. Journ. Fish* **9**, 213–431.

KRUMBEIN, W. C. and PETTI, JOHN (1938). *Manual of sedimentary petrography*, 549, Appleton Centuary Crofts Inc. New York.

KURIAN, C. V. (1953). A preliminary survey of the bottom fauna and bottom deposits of the Travancore coast within the 15 fathom line. *Proc. Nat. Inst. Sc. India* **19**, 6, 746–775.

KURIAN, C. V. (1966). Studies on the benthos of the South West Coast of India. Proceedings of Symposium on Indian Ocean. *Bull. Nat. Inst. Sc. India* **38**, 3.

LONGHURST, A. R. (1957). Density of Marine benthic communities off West Africa. *Nature Lond.* **179**, 542–543.

MARE, M. F. (1942). A study of a marine benthic community with special reference to the microorganism. *J. Mar. Biol. Ass. U. K.* **25**, 517–554.

McINTYRE, A. D. (1968). The meiofauna and macrofauna of some tropical beaches. *Journ. Zoology* **156**, 3, 377–393.

McINTYRE, A. D. (1969). Ecology of Marine Meiobenthos. *Biol. Rev.* **44**, 245–290.

MORTENSEN, T. (1925). An apparatus for catching the microfauna of the sea bottom. *Vidensk. Medd. densk. naturh. Foren Kbh.* **80**, 445–451.

MUUS, B. J. (1964). A new quantitative sampler for the meiobenthos. *Opheiia*, **1**, 209–216.

PETERSEN, C. G. J. and JOHNSON, P. B. (1911). Valuation of the sea 1. Animal life of the sea bottom, its food and quantity. *Rep. Danish Biol. Sta.* **20**, 3–79.

PETERSEN, C. G. J. (1915). A preliminary result of the investigations on the valuation of the sea. *Rep. Danish biol. Sta.* **23**, 29–32.

REES, C. B. (1940). A preliminary study of the ecology of a mud flat. *J. Mar. Biol. Ass. U. K.* **24**, 185–199.

REMANE, A. (1933). Verteilung und Organisation der benthonischen Mikrofauna der Kieler Bucht. *Wiss. Meeresunters.* (Abt. Kiel) **21**, 161–221.

RAO, K. V. (1969). Distribution pattern of the major exploited marine fishery resources of India. *Bull. Cent. mar. Fish. Res. Inst.* **6**, 1–69.

SESHAPPA, G. (1953). Phosphate content of mud banks along Malabar coast. *Nature Lond.* **171**, 526–527.

SESHAPPA, G. (1953). Observations on the physical and biological features of the inshore sea bottom along the Malaber coast. *Proc. Nat. Inst. Sc. India* **19**, 2, 257–279.

SESHAPPA, G. and JAYARAMAN, R. (1956). Observations on the composition of bottom muds in relation to the phosphate cycle in the inshore waters of the Malabar coast. *Proc. Indian Acad. Sci.* **43**, 288–301.

SILAS, E. G. (1969). Exploratory fishing by R. V. Varuna. *Bull. Cent. Mar. Fish. Res. Inst.* **12**, 1–86.

SMITH, W. and McINTYRE, A. D. (1954). A spring loaded bottom sampler. *J. Mar. Biol. Ass. U. K.* **33**, 257–264.

SPARCK, R. (1935). On the importance of quantitative investigation of bottom fauna in marine biology *J. Cons. Inst. Explor. Mer.* **10**, 1, 3–19.

STEPHEN, A. C. (1923). Preliminary survey of the Scottish waters of the North Sea by the Petersen Grab. *Rep. Fish. Bd. Scot.* 1922, **3**, 1–21.

THORSON, G. (1957). Bottom communities (sublittoral and shallow shelf) *Mem. geol. Soc. Amer. Mem.* **67**, 1, 461–534.

Oceanography and its relation to marine organic production

E. C. LAFOND and K. G. LAFOND

Naval Undersea Center, San Diego, California 92132, U.S.A.

Resumo

A produção orgânica no Oceano é o resultado de processos complexos e interrelacionados. Para um dado organismo desenvolver-se são necessários requisitos nutricionais especiais, clima biológico e condições físicas tais como, luz, temperatura e movimentos da água. São discutidos êsses aspectos da oceanografia, relacionados com o ciclo de vida orgânica, no oceano, e os fatôres físicos do ambiente marinho, os quais influenciam o movimento e distribuição das propriedades da áqua.

1. A topografia circunjacente incluindo configurações costeiras tais como ilhas, promontórios e "canyons", e outras variações em profundidade, indluenciam a circulação vertical e horizontal no Oceano.

2. Os limites das principais correntes oceânicas, criam movimentos da água circunjacente. Êsses limites consistem principalmente de "eddies" e frentes, os quais contém convergâncias e divergências que criam ambiente favorável para a vida marinha.

3. Oscilações verticais comumente chamadas ondas internas estão presentes em todos os oceanos. Tais mudanças cíclicas no nível das partículas na coluna de água, causam variações no conteúdo, luz, e pressão, que podem ser benéficos para produção orgânica.

4. A microestrutura, ou formação de camadas no Oceano, delineia o fluxo e as propriedades da água. Camadas muito finas, frequentemente apresentam descontinuidades que indicam fluxo diferencial. Organismos planctônicos e larvas podem ser dispersos em muitas direções por êsses fluxos. Tais camadas criam grandes possibilidades nas condições ambientais, algumas das quais podem ser ótimas para os requisitos de reprodução

dos organismos. Apesar dos programas oceanográficos dos quais êste estudo foi derivado, terem sido realizados em regiões oceânicas diversas daquelas próximas às Américas Central e Sul, a técnica desenvolvida e a informação adquirida, podem ser aplicadas a futuros programas a serem projectados para a América Latina.

INTRODUCTION

For the sea to be fertile and yield its abundance of food, a sequence of processes and properties must exist. First there is the energy from the sun. Then there is the motion of water, which affects the distribution of a variety of chemicals found in the sea. The properties of light and topography, pressure and temperature, accompanied by constant motion, produce environments fit the propagation of living organisms, each one adapted to a specific environment by prior natural selection.

Any understanding of the nutritional enrichment of the oceans must begin with a systematized oceanograhpic approach. This discipline furnishes the necessary knowledge of the specialized physical, chemical, and biological processes within a given area that favor the production of plankton, for it is plankton growth that makes possible the larger life forms that can provide an ever-multiplying humanity with food.

The organic cycle begins with the presence in the euphotic zone of living phytoplankton cells in contact with nutrient-rich water. Vertical and horizontal motions redistribute plant nutrients and create a variety of conditions favorable to the growth of particular organisms. These motions include waves, convection circulations, upwelling, eddy circulation and diffusion, and tidal flows. The surrounding topography including coastal configurations and spatial variations in water depth, also influences the vertical and horizontal circulations. Measured vertical circulations show largescale flows in water-mass boundaries. Internal waves, always present in the thermocline, mix the water and transport living organisms. The laminar layering of partially mixed waters in the thermocline tends to move water horizontally in different directions. Oceanographers have established that one of the most effective circulations for the support of food production occurs in regions where the phenomenon of upwelling takes place. It is possible that observations from outer space (Fig. 1), which encompass vast oceanic areas may yield, through cloud patterns and sea surface features, evidence of the locations of such productive regions in the world's watery expanse. This report presents an overview of some of the physical factors known to be critical for the initiation and support of life in the sea and emphasizes the various water motions which are so important to these processes.

THE BASIS OF FERTILITY

For the sea to be fertile and life to flourish, a number of environmental conditions are necessary, including biological climate, nutritional, and physical factors (Sverdrup *et al.*, 1942; Moore, 1958; Nicol, 1960; and Raymont, 1963). The principal nutritional need is the variety of chemicals that make up the molecules of living matter and make possible their growth. The physical factors in the sea include such properties as light, topography, temperature, pressure, and motion. Each organism requires a special combination of environmental conditions. One species could thrive in a climate that would be fatal to another. Biologically supporting associations must exist between one organism and the others. Although certain mechanisms may create high-nutrient areas, little production of organic matter can proceed unless the total environment is suitable for living organisms. Furthermore, the ocean must be "seeded" with the organisms before growth can occur.

Life Cycle

Photosynthesis is the major producer of organic matter in the sea, but essential plant nutrients, chiefly phosphates, nitrates, silicates, and trace elements such as iron and manganese, and some organic compounds or so-called accessory growth factors must first be present (Fig. 2). Under the influence of sunlight, these nutrient salts are utilized by marine plants for the production of organic compounds.

Primary production takes place through the photosynthetic activity of the microscopic phytoplankton, part of which is subsequently consumed by zooplankton. The smaller marine animals in turn serve as food for large carnivores. An inefficient conversion of energy and inhospitable environment cause losses in organic matter to occur all along the food chain. The weight of the large carnivores may be less than 1/1000th that of the phytoplankton mass. The organic cycle is finally completed when the nutrients locked in the plant and animal tissue and the products of metabolism are returned to the sea through bacterial decomposition.

16*

Movement

The simplified cycle shown in Figure 2 is complicated by many factors. The movements of organisms into or out of a particular area will alter the local balance, and measurements of the processes pertinent to rates of production and mortality are needed to estimate the long-term capacity of the ocean to produce biomass at each trophic level. The standing crop can be maintained at a uniform size either by a small-scale cycle with a high cycling rate or by a large-scale cycle with a low cycling rate. The time taken for a cycle differs considerably for various organisms.

Light

The type and number of organisms constituting the standing crop are strongly affected by environmental influences, among which light is a major element. The intensity of incident light impinging on the sea surface varies with latitude, season, time of day, and amount of cloud cover. The amount of light that penetrates the sea depends on surface reflection, which changes with the altitude of the sun, sea surface roughness, and attenuation in the water.

The rapid attenuation of the usable wavelengths of light in the water causes only the near-surface zone to be illuminated enough to allow photosynthesis and primary production of organic matter. Generally, the euphotic layer is thickest in low latitudes and thinnest toward shore. In bright sunlight the uppermost part of the layer may be too intensely illuminated to permit maximum photosynthesis. The optimal depth may lie somewhere between the surface and a depth of 20 to 30 meters (Nielsen, 1955). Below this the rate of photosynthesis appears to be directly proportional to light intensity.

Light plays an important part in the behavior of phototrophic organisms, especially those that make up the migratory deep-scattering layers (Kampa and Boden, 1954; Clark and Backus, 1956; and Barham, 1966). Seemingly unhindered by temperature and density barriers, the organisms accompanied by predators, migrate hundreds of meters daily.

Topography

A strong influence in promoting organic production is proximity of the organisms to the sea bottom or shore. In addition to the introduction of many nutrients through the erosive action of rivers, tides and currents, the near-shore zones are requisite for spawning, feeding and protection of certain species of marine life. The type of sediment and changes in water depth are thus important. The sea bottom topography, however, can provide barriers that limit the spread of benthic organisms; also some species thrive outside the range of terrigenous influence.

Temperature

A correlation exists between temperature and the distribution of marine organisms (Hubbs, 1948). Each species propagates best within a certain temperature range. Since the sea temperature influences the chemical processes, respiration, and metabolism of organisms, it is of prime importance to distribution. Temperature, however, is largely controlled by weather, and the thermal structure of the water column is influenced by various subsurface motions. Inversely, the density structure restricts or promotes the vertical, upward transfer of vital nutrients (Armstrong and LaFond, 1966).

Water motion, a prime study of oceanographic science, thus becomes an important link in the investigation of marine organic production.

WATER MOTION

Major, as well as small scale water motions in the sea can exert either beneficial or adverse effects on marine life. They can, for example, allow organisms to utilize differential current systems and perform limited horizontal navigation, or expedite the vertical transport of dissolved and particulate substances. Diverging currents displace organisms (Moore, 1962) to favorable or unfavorable environments; i.e. introduce a supply of an initial stock of suitable organisms or dilute the population of the euphotic layer. Mixing circulations can cause changes in the temperature, chemical content, light intensity, or pressure in a given volume of seawater containing the organisms. Mixing processes can move food to a location where it can be utilized by predators, or they can be instrumental in removing toxic waste products.

The nutrient requirements of marine plants are similar to those of land plants. The nutrients and accessory growth factors are present in the sea in small quantities and must exist in the euphotic zone to be utilized by the phytoplankton. The mechanisms of transport and mixing by which the surface layers are replenished with nutrients are therefore of prime importance in productivity. The near-surface influences include river runoff, turbulence and eddies, convection currents, wind-and-wave mixing, cascading and capsizing, tidal circulation, upwelling, internal waves, and shoal effects (LaFond, 1954, 1963), (Fig. 3).

River runoff

Fresh, nutrient-laden river water containing dissolved minerals and decomposed terrestrial matter leached from land, spreads out in the oceans at the mouths of rivers, mixes with seawater, and moves with the currents (Fig. 3 A). Only part of the added nutrient material, however, remains in the suface layers; the remainder is soon mixed and eventually distributed in deeper layers. As a result of river runoff and mixing at the confluences, coastal water exhibits a wider variety of properties, but it generally contains more dissolved oxygen, and organic substances, and commonly supports a higher concentration of plankton, benthos, bacteria, and fish, than offshore water. A main source of essential oxygen in the near-surface waters comes from the action of breaking waves created by strong winds. Nearshore waters have been found to be particularly productive along the boundary, or mixing region, between the two water types. The chemical and physical properties of the mixing waters frequently develop foam. The boundary areas, characterized by strong salinity gradients, can prevent the free onshore-offshore movement of species adapted to coastal or oceanic water. A notable example is the Amazon River outflow where striking spatial differences in planktonic flora are observed (Hulburt and Corwin, 1969).

Turbulence and eddies

In the open sea, most of the plant nutrients are present at levels below the euphotic zone. An important part of food-cycle investigation is to find out how the nutrients reach the near-surface layers where they can be utilized. One way may be by slow molecular diffusion, but a more effective means is through turbulence, which gives rise to eddy diffusion. Where the waters at different depths are of the same density, the process can occur freely; stratification of the water, however, reduces turbulence and eddy diffusion. Thus strong permanent and seasonal thermoclines place a formidable obstacle to the upward transport of nutrients.

Boundaries separating two different water types are generally associated with a differential in the speed and direction of currents, and usually contain eddies. In the oceans north of the equator, cyclonic eddies bring up deeper water and form domes of colder water (Fig. 3B). Such eddies are often favorable areas for organic production because of their higher nutrient content. In the anticyclonic eddies, surface waters accumulate and sink, whereupon they not only introduce oxygen-rich water to greater depths but entrap surface film and debris. This in turn attracts a variety of small organisms. The two types of circulation which are sometimes found together may therefore be favorable for the growth and concentration of plankton and subsequent attraction of fishes.

Convection currents

A special mixing process occurs when certain meteorological conditions reduce surface temperature or increase evaporation. This causes the uppermost sea layer to become denser than those immediately below. The denser water then sinks and produces a convective overturn (Fig. 3C). Convection cells are thus set up in which subsurface water, often relatively rich in nutrients, rises and mixes with heavier, sinking water.

Wind-and-wave mixing

The wave-creating, mechanical action of wind on the sea surface is an important force in mixing the upper layers and bringing nutrients from the thermocline into the euphotic layer. Wind waves and swells created by local and distant storms cause the surface water to undergo orbital motion in a vertical plane normal to the direction of wave propagation (Fig. 3D). The orbital motion decreases with depth to about 4 percent of its amplitude at half a wavelength (Defant, 1961). Wave mixing is most effective when the waves are long and the thermocline is shallow, and in shallow water where the wave motion extends to the bottom.

Strong winds create an additional surface-mixing system by setting up long wind-driven, vertical convection cells (Langmuir, 1938; and Woodcock, 1944), the axes of which are oriented at a small angle to the wind direction (Faller, 1964).

Cascading and capsizing

A phenomenon called "cascading", was investigated (Cooper and Vaux, 1949) off the continental slope of the southwest British coast. It is believed that the marked winter cooling of the sea surface leads to such mass sinking that a flow of dense water is created near the bottom at the edge of the continental slope. The flow of deep cascading water along submarine valleys might have an effect on deeper currents.

Cascading would deplete coastal areas of nutrients. However, a second concept called "capsizing", would enrich them. Capsizing occurs when the surface water is turned below as the result of instability caused by continued, strong onshore winds, and denser, deeper water is brought to the surface. The water mass thus quickly becomes homogenously mixed.

Tidal circulation

Tidal currents, caused by periodic changes in the gravitational attraction of the moon and sun, set up oscillations of known frequencies; the major ones being of 12.4 and 23.9 hours. Since the natural oscillation frequency of the larger oceans is nearly equal to that of the periodic attraction of the sun and moon, large water displacements result. The movement of the water produces significant currents, especially in shallow waters and narrow sounds (Fig. 3E). The currents impinge on submerged shoals and peaks and to a lesser extent on the deep sea floor, and are effective in mixing as well as transporting water. The action in flushing harbors, river mouths, and estuaries causes major changes in the chemical and biological properties of the environment, and the nutrient potential is altered.

Upwelling

Of the water motion processes, one of the most important to organic production is upwelling. Although the area of the vertical motion may be geographically small, the upwelled water and its effects on sea life can extend for hundreds of miles. Upwelling may take place anywhere, but is most prevalent along the western coasts of continents, notably Peru, United States, Morocco and South Africa. An extensive area also occurs off the coasts of Somaliland and Arabia (LaFond, 1966a).

Upwelling may be caused by wind displacement of surface water from the coast or by diverging currents. At the equator, upwelling is caused by a divergence related to a reversal of the Coriolis Force. Where adjacent surface waters flow away from each other, the deeper water must rise. In addition, large and small cyclonic gyrals can force displacements upward and produce upwelling.

Most regions of upwelling are associated with coastal wind-drift currents. When the wind blows nearly parallel with the coast in the northern hemisphere (with the water on the right-hand side, facing downwind), the surface water will be displaced offshore in accordance with Ekman transport. To take its place, subsurface water upwells near shore and probably along upward sloping density boundaries. The process is defined as upwelling whether or not the upwelled water physically reaches the surface. The degree of upwelling depends on the angle of the wind in relation to shore configuration, and on wind strength and duration; consequently it varies with the season. Coastal upwelling is probably the most effective means by which high-nutrient water is raised to the surface layer (Fig. 3F).

Internal waves

A feature in water mixing, which has an attendant effect on marine organic production through vertical and horizontal displacement of particles, is the phenomenon of internal waves (LaFond, 1966b). These waves are evidenced by repeated vertical changes in isotherm depths (Fig. 3G). Internal waves are found between layers of different density or within layers where different density gradients are present. This is usually a nutrient boundary as well.

Internal waves have been examined by analysis of their influence on sea properties and by direct measurement, notably from the Oceanographic Research Tower off Mission Beach, California. They have an amplitude at all depths except at the sea bottom, where it is zero, and at the free surface, where it is negligibly small. When the waves are near the sea surface their convergence circulation is capable of concentrating surface film and producing slicks. Water motion over the crest may even ripple the sea surface. When an internal wave thermal boundary is near the sea floor, a similar action exists. The maximum turbulence will be under the trough, but the maximum speed will be in the direction opposite to the wave propagation. The funneling of water through the constriction created by the trough and bottom, always in an offshore direction, is undoubtedly a contributor to the offshore movement of sediment and stirring of detritus from the sea floor. Even in deep water, internal waves near the bottom can move sediment and form ripple marks, as well as mix nutrient-rich water (LaFond, 1961). Higher internal waves frequently occur in groups and exert a profound effect on the movement of water properties and life within the sea (LaFond, 1962a, b).

BOUNDARIES

Shoals

The direction and speed of water flow are influenced by the depth and configuration of the coastline. The distribution of land and water creates a differential in heating, a major cause of winds, which in turn produce water circulations.

Surface waves moving towards shore couple with edge waves moving along shore to create zones of high and low water level that develop rip currents, spaced at the wavelength of the edge waves. These transport plumes of coastal water seaward which eventually mix with offshore water.

When an island is in the path of a current, surface water tends to accumulate on the up-current side, and a localized upwelling occurs on the down-current side (Fig. 3 H). A current passing an island creates turbulence along its sides and large, transient eddies in the down-current wakes. Off Hawaii, several topographic features simultaneously influence the thermal structure (Fig. 4A). Turbulence, and a ridge effect, are present south of the island of Hawaii and along its sides. In the lee of Hawaii and Maui Islands a large, counterclockwise eddy repeatedly forms and causes divergence and upwelling. A greater organic production is observed near islands than in offshore areas and is attributed to the island-mass effect (Doty and Oguri, 1956).

In addition to islands, promontories produce similar eddies on their lee sides. Depending on the direction of current, orientation of the relative shore configuration, and latitude (N or S), the eddy will cause water to rise or sink. In the northern hemisphere, the eddies with counterclockwise circulation create domes of colder water that cool both the subsurface as well as the surface water (Fig. 4B).

Off Baja California, México, two points, Punta Banda and Punta Baja project into a southeasterly flowing current. As a result of upwelling, the surface water temperature south of the points averages 4.4°C and 0.7°C colder, respectively, than that to the north (Dawson, 1951; Hubbs and Hubbs, 1960).

The interposition of shoaling topography in a current is analogous to air flow over mountains (Lyra, 1943) and causes vertical water motion, turbulence, and horizontal eddies that help to maintain surface water in a nutrient-enriched state (Fig. 4C). The thermal layers in the vicinity of the shoals rise and create internal waves of a type different from those in adjacent areas. The sediments on the shoals are usually coarse and afford a

Figure 1 The earth from space shows South America and the Atlantic Ocean with cloud patterns related to the marine environment

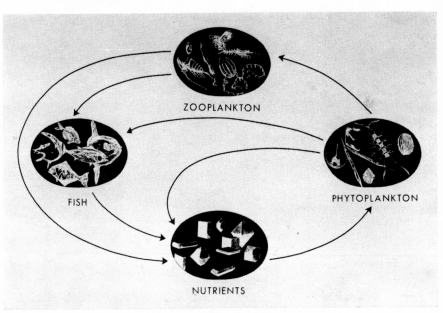

Figure 2 Life cycle in the sea

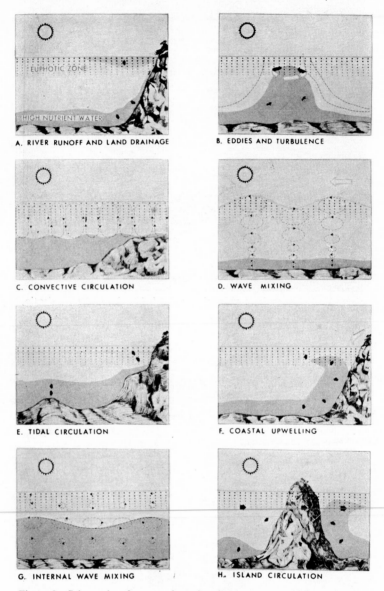

A. RIVER RUNOFF AND LAND DRAINAGE

EUPHOTIC ZONE

HIGH NUTRIENT WATER

B. EDDIES AND TURBULENCE

C. CONVECTIVE CIRCULATION

D. WAVE MIXING

E. TIDAL CIRCULATION

F. COASTAL UPWELLING

G. INTERNAL WAVE MIXING

H. ISLAND CIRCULATION

Figure 3 Schematic of types of motion in the ocean which can enrich
the upper layers of the sea

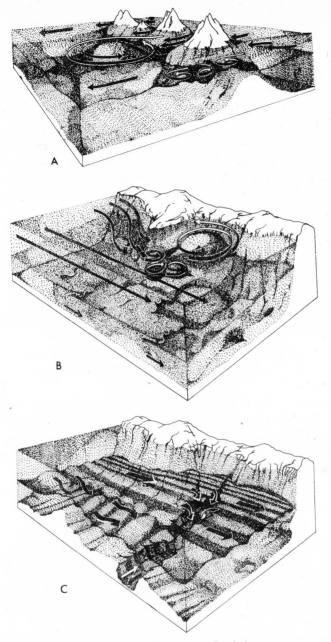

Figure 4 Land influences on circulation

A   The southern Hawaiian Islands create a large eddy in their southwest
      lee and smaller scale turbulence off the southern coast

B   Land promontories along the Baja California Coast can cause eddies
      and upwelled colder water on the lee sides

C   Shallow water topographic features such as ridges and submarine
      canyons influence coastal circulation and transport

favorable accommodation for a variety of life whose presence can be detected by acoustic means.

The continental slope and shelf are other topographic features that afford a variety of environments for the support of marine life (Shepard and Dill, 1966). As tidal currents or internal waves impinge on the sea floor, they are deflected and stir up sediment and enrich the water column.

The funneling of water into and out of submarine canyons has been measured (Shepard and Marshall, 1969), and visual observations from submersibles have noted that the fauna is more abundant in the canyons than in the adjacent areas.

Water masses

The oceans have been divided according to their temperature-salinity relationships into large-scale identifiable water masses, hundreds of miles in extent, which have distinguishable properties (Fig. 5). Thermal fronts observed in the boundary regions between the water masses show marked gradients in temperature, salinity, and other properties. The temperatures and relative current speeds in and around these boundaries have been measured with a towed oceanographic thermistor chain (LaFond and LaFond, 1967) (Fig. 6A).

The significant data for vertical motion are the relative differences in the speed of the flow at the various levels as the chain is towed through the water (Fig. 6B). By using the near-surface meter (13 m) as a reference and measuring the difference between the recording of the surface meter and the

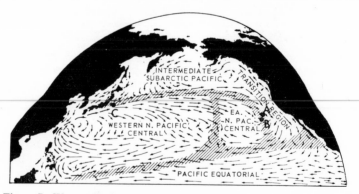

Figure 5 Water mass boundaries in the North Pacific Ocean are indicated by shading. Small arrows represent surface current. Examples of temperature and current data are from positions A, B and C

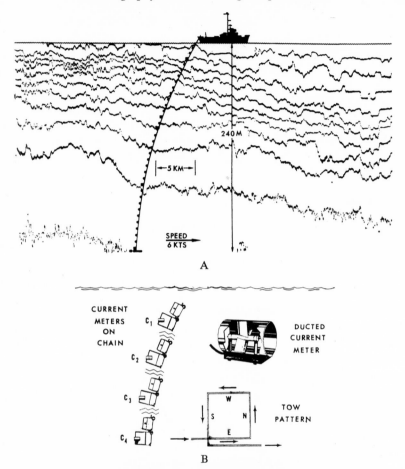

Figure 6 A   Vertical temperature structure recorded by towed thermistor chain. The 2-D isotherm patterns (in 1°C intervals) are automatically recorded in real time

B   Ducted current meters mounted at four levels along the chain are used to determine relative current simultaneously with temperature structure

other meters (e.g. C_1–C_2) the horizontal shear can be established in a given plane. Some clue as to the direction of flow is furnished by the relative level of isotherms recorded simultaneously with the current. Although different interpretations can be made of these data, they do demonstrate that much vertical circulation is present in the upper 240 meters of the water column, a motion that will displace organisms and their food to different levels.

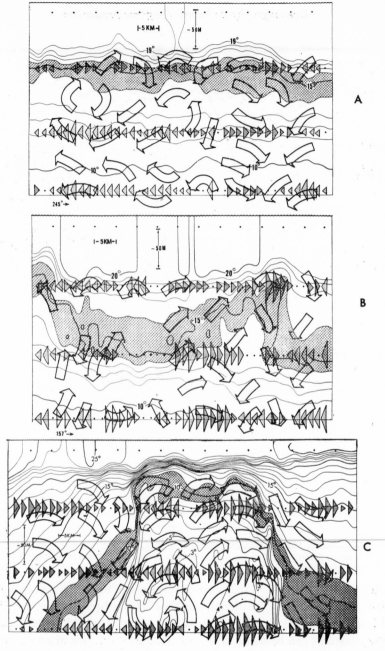

Figure 7 Vertical temperature structure in isotherm intervals of 1°C. Horizontal rows of triangular arrows show the relative current speed and direction between the near surface meter and indicated depths. For each section the speed is proportional to the size of the arrow. The large, open arrows show the circulation as interpreted from the data

A   Data from position A in figure 5. Maximum relative current is 0.6 knot
B   Data from position B in figure 5. Maximum relative current is 0.4 knot
C   Data from position C in figure 5. Maximum relative current is 0.8 knot

Three examples of vertical sections (Figs. 7A, B and C) in frontal zones are presented. The probable relative vertical and horizontal circulations are deduced from the temperature recorded simultaneously with the current as it moves past the towed thermistor chain. In most of the oceans, the horizontal temperature structure is relatively flat. However, at the large water-mass boundaries, which are made up of sharp fronts, major horizontal and vertical changes exist in isotherm depths (LaFond, 1968).

The movement and redistribution of water masses and their contents are beneficial for organic production and fisheries (Uda, 1952, 1968; Uda *et al.*, 1958). It is probable that throughout the world, as a result of these mixings, in the large ocean areas there is such a wide range of water properties that at some level an appropriate set of conditions will be found to be optimum for a particular organism, and there it will develop at an accelerated rate at this position in time and space.

Slicks

Sea surface slicks, which are often visible evidence of internal-wave circulation, appear on a relatively calm surface as streaks or patches surrounded by rippled water (LaFond, 1969). The visibility of slicks is contingent upon wind, lighting, presence of sufficient organic matter on the surface, and the nature of progessive-type internal waves, such as their height, depth, and length.

Surface slicks are nearly always found directly above a descending thermocline (Fig. 8) somewhere between the crest and the following trough. The significant motion is a surface convergence that causes a slick to be formed over the trailing slope of an internal wave. This is the result of water circulation created by progessive-type internal waves. As the internal waves propagate shoreward across a continental shelf, the slicks retain the same relative position to the wave, i.e. just behind the crest. Thus by noting the position and motion of sea-surface slicks, the position and motion of internal waves can be determined.

Analysis of the major water-insoluble organic constituents of slick material, collected in wide areas of the open sea, showed them to be complex mixtures of fatty esters, free fatty acids, fatty alcohols, and hydrocarbons (Garrett, 1967a). In addition, Hoyt (1968) found that algae and diatom cultures secrete, in the process of metabolism, polysaccharides of large molecular weights. The most likely sources of this material are the metabolic products

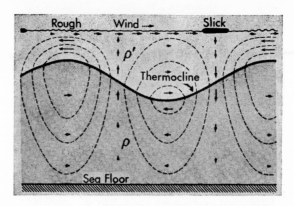

Figure 8 Circulation associated with progressive-type internal waves in a twolayer density (ϱ', ϱ) system. Fine dashes are streamlines. The average position of sea surface slicks to internal wave phase is indicated

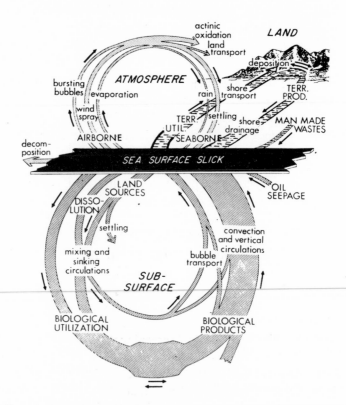

Figure 9 Schematic of sources and processes in the cycles of sea surface slick film. The thickness of lines estimates the importance

and decomposition of organic matter in the sea, which provide substances for further organic production.

Cycles in the surface-active, slick-producing material are shown schematically in Figure 9 and apply to the near-shore area studied off Mission Beach, California. An estimate of the importance of the sources and processes is expressed in the thickness of the bands.

Land

Shore drainage can introduce large quantities of decomposed, terrestrial organic material. Other substances are added by man through sewage discharge and industrial wastes. In this category can be included the oil pumped and leaked from vessels and the natural, subterranean oil seepage in certain areas.

The main process by which near-shore film is returned to land is by direct transport on surface water. The circulations associated with internal wave propagation and wind move the slick bands toward shore. The speed of slicks was measured at roughly 0.5 knot when moving past the Mission Beach, California, oceanographic tower site. They covered approximately 14 per cent of the sea surface and if their thickness is 30 Å, the volume would amount to about 9 ml/meter of coastline per day. Some of the material is probably broken down in the surf zone by breaking-wave action and dispersed into the atmosphere. However, a major quantity of the organic material is deposited on the beach, where it affords sustenance for life on the shore and in interstitial regions. This material is washed and leached by waves and tides. Some of it refloats on the sea surface; some enters the subsurface and, to a lesser extent, the atmospheric cycle. That on the surface is subjected to buffeting by both water and air, and changes, including decomposition (ZoBell, 1962), take place through photocatalyzed oxidation, which loses material to the cycle.

Atmosphere

Some airborne organic particles, orginating mainly from the sea, resettle on the sea surface. The settling is aided by vertical wind components and especially through the action of rain.

The processes causing these particles to be airborne from the sea surface are evaporation, bubble bursting (Abe, 1962), and wind spray (Blanchard, 1964). The partially volatile portions of surface material gradually evaporate. A significant atmospheric source, however, is the bursting of bubbles at the sea surface. Bubbles are produced both within the water column, and at the surface by wind breaking the ridge of waves.

A high percentage of bubbles normally accumulates in the slick bands (LaFond and Dill, 1957). In the water column, bubbles are produced by the heating of gas-saturated water (Ramsey, 1962a, b), which frequently occurs in the spring and summer. *In-situ* bubbles can also be produced through a vertical change in the level; this reduces the pressure of gas-saturated water. Both air and oxygen can be released into the atmosphere, but they also carry with them water particles, plant nutrients (Wilson, 1959); microorganisms (ZoBell, 1942); salt particles (Blanchard, 1963 and Day, 1964); organic particles (Sutcliffe *et al.*, 1963; Blanchard, 1964); and slick film substances (Garrett, 1967b). Thus through the action of bubbles, especially in slicks where so many accumulate, an important process occurs in the transporting of water properties and the associated movement of nutrients.

Subsurface

The greatest quantity of organic material comes from the marine biosphere. In the metabolism and decomposition of vast quantities of organisms in the sea, the insoluble parts that are lighter than water must rise to the surface. Some organic particles become attached to bubbles, which aid them in their upward transport.

The downward displacement of film into the subsurface water takes place in mixing and sinking circulations. Some material dissolves in the underlying water and is utilized by organisms. It is estimated that one-third of the hydrocarbons may be converted into bacterial cells. These in turn can be utilized by microorganisms, which are themselves ingested by larger organisms to ultimately proceed through the entire organic web in a never-ending cycle. Of significance to this cycle is the fact that organic matter is recycled and can add to the fertility of the sea.

From manned submersibles, small particles resembling snow have been observed throughout the entire water column and are believed to be decomposing organic matter. When organisms die, the heavier components tend to settle to the sea floor where they are utilized by benthic organisms. Near-sea-floor sampling at 1300 meters indicates an absence of distinct nutrient gradients (phosphate, nitrate, silicate, nitrite) within 10 meters of the sea floor. This may be due to their rapid utilization by sea bottom organisms and to the action of convection currents caused by instability of bottom water as a result of sea floor heat flow. Such circulations tend to mix the lower level and redistribute nutrients in the water column.

STRATIFICATION

Microstructure

Both turbulence and stratification of the water column have beneficial as well as adverse biological significance. Vertical casts made by sensitive instruments have shown that the thermocline in the ocean is made up of layers (Lovett, 1963; Holm–Hansen *et al.*, 1966; Cooper and Stommel, 1968), some at least as thin as a few centimeters (Cox *et al.*, 1968). Many of these microstructures show geographical differences as well as characteristic sharp delta depth patterns having differential flows (Piip, 1968; Wennekens, 1968; and Lovett, 1968).

Some revealing information on the nature of the motion of thin layers comes from dye studies made in clear Mediterranean water (Woods, 1968a, b). From the sharp horizontal changes of dye patterns it was found that thin layers of constant temperature, separated by sharp gradient layers 10–20 cm thick, called sheets, can be clearly delineated in the thermocline. Turbulence is greatly suppressed in the thin sheets with lowered viscosity and they lend themselves to smooth laminar flow in widely varying directions.

These layers undergo spatial and temporal depth changes in accordance with the thermocline. One example of horizontal changes is the record of the depth of whole degree isotherms, made off southern California by the towed temperature sensors on the thermistor chain (Fig. 10). The isotherms make vertical excursions of varying amounts, up to 15 meters over a distance of 2 kilometers. In this section the isotherms are not all in phase with each other, indicative of subsurface advection. Another significant feature is that the repeated recordings of the depth of all isotherms (11–16°C) show that their level fluctuated a meter or two in a sampling distance of 37 meters. Such variation is further evidence that these layers must move up and down with internal waves and other subsurface currents.

Stratified layers have been studied for biological activity and found to contain considerable differences in the organisms present (Strickland, 1968). One oceanographer reported seeing a thin layer of organisms above him as he looked up from under water. The layers help to conserve energy for the organisms since they can be transported by the flow of water for long distances without swimming, but an expenditure of considerable energy is

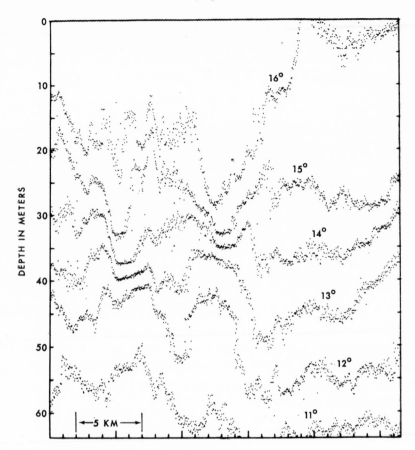

Figure 10 Vertical temperature structure on expanded scale determined
with the towed thermistor chain. Recorder scans, indicated by vertical
series of fine dots, each representing the depth of a whole degree centigrade,
are made at 12-second intervals; e.g., 37 m when travelling at 6 knots.
Note the horizontal variation in depth of isotherms

required for the organisms to move up and down through the layers. This
was shown by sonic scatter records (Southward, 1962), which revealed a
variation in depth of a layer of scatterers, attributed to internal waves in the
thermocline. One interpretation is that the scatter was caused by a congrega-
tion of animals undergoing vertical migration at night, but who on this
occasion were unable or unwilling to pass through the boundary layer into
the body of very much warmer water above. The work of Angel (1968) also

showed that the thermocline acts as an ecological boundary for many plank-tonic species. Thus the complex water motions associated with layered thermoclines play an important part in dispersing both nutrients and organisms in the sea.

SUMMARY

Organic production in the sea is the result of complicated, interrelated processes. For a given organism to flourish, special nutritional requirements, biological climate and physical conditions, such as light, temperature, and water motion, must exist. Those aspects of oceanography pertinent to the organic life cycle in the ocean and the physical factors of the marine envi-ronment which, through their influence on water motion and the distribution of water properties, help make that life cycle possible have been discussed.

1. The surrounding topography, including coastal configurations such as islands, shoals, promontories and canyons, and other spatial varia-tions in water depths, influences the vertical and horizontal circulation of the ocean.

2. The boundaries of the major ocean currents create relative motion with the bordering water. These water-mass boundaries consist of eddies and fronts that contain ever-changing convergences and divergences which may create favorable environments for marine life.

3. Vertical oscillations, commonly called internal waves, are present in all oceans. Such cyclic changes in the level of particles in the water column cause changes in content, light, and pressure which can be beneficial for organic production.

4. The micro-structure, or layering of the ocean, delineates the flow and the properties of water. Thin, lens-like layers, as narrow as a few centimeters, frequently show sharp discontinuities indicative of differential flow. Planktonic organisms and larva can be dispersed in several directions by these differen-tial flows. Such layers create a wide range of environmental conditions, some of which may be optimum for the reproductive requirements of specific organisms.

Although the oceanographic programs from which this knowledge of the complex fertility of the sea have been derived were largely conducted in oceanic regions other than those bordering Central and South America, it should be emphasized that the technology developed and the information gained are all applicable to future programs projected for the waters of Latin America.

Acknowledgement

We wish to thank E. G. Barham, D. V. Subba Rao and O. S. Lee for helpful suggestions to this manuscript.

References

ABE, T. (1962). On the stable foam formation of sea water in seas (preliminary report). *J. Oceanographical Society of Japan, 20th Anniversary Volume.*

ANGEL, M. V. (1968). The thermocline as an ecological boundary. *Sarsia,* 34, 299–312. 2nd European Symposium on marine biology.

ARMSTRONG, F. A. J., and LAFOND, E. C. (1966). Chemical nutrient concentrations and their relationship to internal waves and turbidity off southern California. *Limnology and Oceanography.* 11, 4, 538–547.

BARHAM, E. G. (1966). Deep scattering layer migration and composition: observations from a diving saucer. *Science,* 151, 3716, 1399–1403.

BLANCHARD, D. C. (1963). The electrification of the atmosphere by particles from bubbles in the sea. In *Progress in Oceanography* vol. 1, pp. 71–202. (Ed. M. Sears) Pergamon Press, Oxford.

BLANCHARD, D. C. (1964). Sea-to-air transport of surface active material. *Science,* 146, 396–397.

CLARK, G.L., and BACKUS, R.H. (1956). Measurements of light penetration in relation to vertical migration and records of luminescence of deep-sea animals. *Deep-Sea Res.,* 4, 1–14.

COOPER, J. W., and STOMMEL, H. (1968). Regularly spaced steps in the main thermocline near Bermuda. *J. Geoph. Res.,* 73, 18.

COOPER, L. H. N. and VAUX, D. (1949). Cascading over the continental slope of water from the Celtic Sea. *J. Mar. Biol. Ass.,* 28, 719–750.

COX, C., NAGATA, Y., and OSBORN, T. (1968). Oceanic fine structure and internal waves. In *Commemorative Papers for Dr. Michitaka Uda,* Tokyo University of Fisheries, (in press).

DAWSON, E. Y. (1951). A further study of upwelling and associated vegetation along Pacific Baja California, Mexico. *J. Mar. Res.,* 10, 1, 39–58.

DAY, J. A. (1964). Production of droplets and salt nuclei by the bursting of air-bubble films. *Q. Jl. R. met. Soc.,* 90, 72–78.

DEFANT, A. (1961). *Physical Oceanography,* vol. II Pergamon Press.

DOTY, M. S., and OGURI, M. (1956). The island mass effect. *J. Cons. Int. Explor. Mer.,* 22, 33–37.

FALLER, A. J. (1964). The angle of windrows in the ocean. *Tellus,* 16, 363–370.

GARRETT, W. D. (1967a). The organic chemical composition of the ocean surface. *Deep-Sea Res.,* 14, 221–227.

GARRETT, W. D. (1967b). Stabilization of air bubbles at the air-sea interface by surface-active material. *Deep Sea Res.,* 14, 661–672.

HOLM–HANSEN, O., STRICKLAND, J. D. H., and WILLIAMS, P. M. (1966). A detailed analysis of biologically important substances in a profile off Southern California. *Limnology and Oceanography*, **11**, 4, 548–561.

HOYT, J. W. (1968). Microorganisms—their influence on hydrodynamic testing. *Naval Res. Reviews.* May.

HUBBS, C. L. (1948). Changes in the fish fauna of western North America correlated with changes in ocean temperature, *J. Mar. Res.*, **7**, 3, 459–482.

HUBBS, C. L., and HUBBS, L. C. (1960). Shoreline surface watertemperature data between La Jolla, California and Punta Baja, Baja, California, from University of California, Scripps Institution of Oceanography Data Report: Surface water temperatures at shore stations, United States West Coast and Baja, California, 1956–1959. *S10 Reference 60–27.*

HULBERT, E. M., and CORWIN, N. (1969). Influence of the Amazon River outflow on the ecology of the western tropical Atlantic. III, The planktonic flora between the Amazon River and the windward island. *J. Mar. Res.*, **27**, 55–72.

KAMPA, E. M., and BODEN, B. P. (1954). Submarine illumination and the twilight movements of a sonic scattering layer. *Nature*, **174**, 869–871.

LAFOND, E. C. (1954). Factors affecting vertical temperature gradients in the upper layers of the sea. *The Scientific Monthly*, **78**, 4, 243–253.

LAFOND, E. C. (1961). Internal wave motion and its geological significance. In *Mahadevan Volume, A Collection of Geological Papers*, pp. 61–77. Osmana University Press. Hyderabad, India.

LAFOND, E. C. (1962a). Internal waves (Part 1). In *The Sea* **1**, pp. 731–751. Interscience Publishers.

LAFOND, E. C. (1962b). Marine meteorology and its relation to organic production in south-east Asian waters. *J. of the Marine Biological Association of India*, **4**, 1, 1–9.

LAFOND, E. C. (1963). Physical oceanography and its relation to the marine organic production in the south China Sea. In *Ecology of the Gulf of Thailand and the South China Sea—A Report of the Results of the NAGA Expedition, 1959—1961.* SIO Reference 63–6, pp. 5–33. University of California, Scripps Institution of Oceanography, La Jolla, California.

LAFOND, E. C. (1966a). Upwelling, In *Encyclopedia of Science and Technology* **14**, pp. 210 to 210a. McGraw-Hill.

LAFOND, E. C. (1966b). Internal Waves. In *Encyclopedia of Oceanography*, pp. 402–408. Reinhold.

LAFOND, E. C. (1968). Detailed temperature and current data sections in and near the Kuroshio Current. *An Oceanographic Data Report for Cooperative Study of the Kuroshio and Adjacent Regions (CSK)*, p. 22. Marine Environment Divison, Naval Undersea Center, San Diego, California 92132.

LAFOND, E. C., and DILL, R. F. (1957). Do inversible bubbles exist in the sea? *Navy Electronics Laboratory Technical Memorandum 259.* Navy Electronics Laboratory, San Diego, California.

LAFOND, E. C., and LAFOND, K. G. (1967). Temperature structure in the upper 240 meters of the sea. In Marine Technology Society, *The New Thrust Seaward; Transactions of the Third Annual MTS Conference and Exhibit, 5–7 June 1967, San Diego, California*, pp. 23–45. Marine Technology Society.

LAFOND, E. C., and LAFOND, K. G. (1969). Perspectives of slicks, streaks and internal waves. In *Commemorative Papers for Dr. Michitaka Uda*. Tokyo University of Fisheries, (in press).

LANGMUIR, I. (1938). Surface motion of water induced by wind. *Science* **87**, 119–123.

LOVETT, J. R. (1963). The SVTP instrument and some applications to oceanography. *Instrument Society of America Trans.* **2**, 216–223.

LOVETT, J. R. (1968). Vertical temperature gradient variations related to current shear and turbulence. *Limnology and Oceanography* **13**, 1, 127–142.

LYRA, G. (1943). Theorie der Stationären Leewellenströmung in freier Atmosphäre. *Zeitschrift für Angewandte Mathematik und Mechanik*, Band 23, Heft 1, 1–28.

MOORE, H. B. (1958). *Marine Ecology*. John Wiley and Sons.

MOORE, H. B. (1962). Behavior of plankton in relation to hydrographic factors, *University of Miami, Institute of Marine Science, ML 62186*.

NICOL, C. J. A. (1960). *The Biology of Marine Animals*. Interscience Publishers.

NIELSEN, S. (1955). Production of organic matter in the oceans. *J. Mar. Res.* **14**, 374–386.

PIIP, A. T. (1968). *Precision sound velocity profiles in the ocean. In v. 4: Canary Islands-Gibraltar-Bay of Biscay sound speed, temperature, etc., (June–July 1965)*. TR 6. Columbia University, Lamont Geological Observatory.

RAMSEY, W. L. (1962a). Bubble growth from dissolved oxygen near the sea surface. *Limnology and Oceanography* **7**, 1–7.

RAMSEY, W. L. (1962b). Dissolved oxygen in shallow near-shore water and its relation to possible bubble formation. *Limnology and Oceanography* **7**, 453–461.

RAYMONT, J. E. G. (1963). Plankton and Productivity in the Oceans. In *International Series of Monographs on Pure and Applied Biology* **18**, p 660. MacMillan.

SHEPARD, F. P., and DILL, R. F. (1966). *Submarine Canyons and Other Sea Valleys*. Rand McNally.

SHEPARD, F. P., and MARSHALL, N. F. (1969). Currents in La Jolla and Scripps submarine canyons, California. *Science* **165**, 177–178.

SOUTHWARD, A. J. (1962). The distribution of some plankton animals in the English Channel and approaches; II surveys with the Gulf III high-speed sampler, 1958–60. *J. Mar. Biol. Ass.* U. K. **42**, 275–375.

STRICKLAND, J. D. H. (1968). A comparison of profiles of nutrient and chlorophyll concentrations taken from discrete depths and by continuous recording. *J. Mar. Res.* **13**, 2, 338–391.

SUTCLIFFE, W. H. JR., BAYLOR, E. R., and MENZEL, D. W. (1963). Sea surface chemistry and Langmuir circulation. *Deep-Sea Res.* **10**, 233–243. Pergamon Press Ltd.

SVERDRUP, H. U., JOHNSON, M. W., and FLEMING, R. H. (1942). *The Oceans*. Prentice-Hall.

UDA, M. (1952). On the relation between the variation of the important fish conditions and the oceanographic conditions in the adjacent waters of Japan. *J. Tokyo University of Fish.* **38**, 3, 363–389.

UDA, M. (1968). Fishery oceanographic studies of frontal eddies and transports associated with the Kuroshio Current System including "subtropical countercurrent." In *Symposium on the Cooperative Study of the Kuroshio and Adjacent Regions (CSK) No. 69* 54–55. FAO, Rome.

UDA, M., and ISHINO, M. (1958). Enrichment pattern resulting from eddy systems in relation to fishing grounds. *J. Tokyo University of Fish.* **44**, 1–2, 105–109.

WENNEKENS, M. P. (1968). Pressure (depth) measurements as related to the measurements of gradients in the ocean. *J. Oce. Tech.* **2**, 2, 49–53.

WILSON, A. T. (1959). Surface of the ocean as a source of air-borne nitrogenous material and other plant nutrients. *Nature, Lond.* **184**, 9–101.

WOODCOCK, A. H. (1944). A theory of surface water motion deduced from the wind induced motion of the *Physalia*. *J. Mar. Res.* **5**, 3, 196–205.

WOODS, J. (1968a). CAT under water. *Weather* **23**, 6, 224–236.

WOODS, J. (1968b). An investigation of some physical processes associated with the vertical flow of heat through the upper ocean. *The Meteorological Magazine* **97**, 1148, 65–72.

ZOBELL, C. E. (1942). Microorganisms in marine air. *Aerobiology, Amer. Assoc. Adv. Sci.* Washington, D. C., publ. no 17, pp. 55–68.

ZOBELL, C. E. (1962). The occurrence, effects and fate of oil polluting the sea. *Int'l Conf. on Water Pollution Res.* 1–27. London: Pergamon Press.

Studies with drift bottles in the region off Cabo Frio

ELLEN FORTLAGE LUEDEMANN

and

NORMAN JOHN ROCK

Instituto Oceanográfico da Universidade de São Paulo, Brazil

Abstract

The results of drift bottle recoveries from 3 cruises of the Instituto Oceanográfico between July 1968 and May 1969 are presented and discussed.

It appears that the results can be described in an integral manner by the combined effects of wind transport, the transport of the Brazil current, and coastal current transport.

An hypothetical model is put forward in order to arrive at these conclusions, and some quantitative estimates of the seasonal variation of the Brazil Current are obtained.

Sumároi

Foram apresentados nesta publicação os resulta dos obtidos através do lançamento de garrafas de deriva em três cruzeiros (Julho 1968, Janeiro de 1969 e Maio 1969) realizados pelo Instituto Oceanográfico na areá de Cabo Frio.

Os resultados foram discutidos considerando as condições meteorólogicas e oceanográficas reinantes durante as operações.

Foi possível determinar, de maneira aproximada a velocidade média da corrente do Brasil na sua variação sa zonal de inverno e verão através de um simples modêlo hipotético.

The Oceanographic Institute of São Paulo has undertaken five oceanographic cruises in the region off Cabo Frio between January 1968 to January 1969 with the oceanographic ship *"Prof. W. Besnard"*. The main purpose was that of studying the seasonal variation of the upwelling in this area, as well as its relation to the Brazil Current. Drift bottles were released over the area in order to contribute information regarding the surface circulation.

The results of the drift bottle recoveries of three of these cruises are discussed: that of July 1968, January 1969 and May 1969. There was some recovery from the January 1968 and July 1969 cruises, but the results were not enough to be significant.

The drift bottles used for the survey were thick glass champagne bottles 33 cm in height, 12 cm in diameter. They were ballasted with sand to float upright with 3 cm of the neck exposed, then sealed with corks and finally water-proofed with red sealing wax. Inside the bottle in a water proof envelope was contained a questionaire (in Spanish, Portuguese, and English), as well as a reward notice (Magliocca and Luedemann, 1969).

Drift cards have been used by the Institute, but were found to be very unsatisfactory for conditions on exposed coasts.

The recoveries are grouped according to areas. The earliest recovery for a particular area is taken as being representative for the group (Luedemann, 1967), and (Neumann, 1968). Distances traveled are taken as the shortest distance between the release point and the point of recovery.

It is worth remembering that the velocity calculated is a mean drift velocity for all conditions along the path taken, and includes the time the bottle spent in circular eddies and also on the beach. Consequently the results are nearly always biased towards lower velocities than those encountered in the principal current system, within which the bottle travelled.

THE CABO FRIO REGION AND SOUTHWEST COAST

The studied area was located between approximately 22° S to 25° S and 40° W to 43° W.

Except for two rocky capes (Cabo Frio and Cabo Buzíos), that are subject to heavy surf for the major part of the year, the study area consisted of extensive sandy beaches, where good recovery might be expected. These beaches extend outside the study area as far west as the 44° W meridian.

From there to Santos (46° 20′ W) the coastline is mainly rocky with only intermittent beaches.

However, from Santos to the south as far as Itapocoróia (26° 50′ S), the coast is a low lying sedimentary plain, where good recovery might also be expected. There exists a small stretch until Florianopolis of rocky coastline, and from there on mainly beach until the River Plate Estuary.

It can be seen that the coastline to the South is divided into bands of poor recovery and good recovery conditions.

South of Cabo Frio the Brazil Current forms a series of eddies. It appears that these may have a significant effect on the drift bottle recoveries.

RESULTS

During the survey off Cabo Frio 1230 drift bottles were released, 10 on each station, out of which 42 bottles were recovered (3.4%). Generally, the release points, from which bottles were successfuly recovered, were located on the continental shelf within 40 nautical miles (n. m.) from the coast.

The results for each cruise are now presented separately and subsequently, comparison, discussion and conclusion of results are considered.

July 1968

Of the 150 bottles launched on the July 1968 cruises, 8.7% or 13 bottles were recovered. Of these, seven were recovered from drift bottle releases within the study area, but from stations located outside the continental shelf (Fig. 1); and six were recovered from stations more to the southwest, near S. Sebãstiao Island (24° S 45° W). Half of the bottles converged to the coast of Sta. Catarina State (between S. Francisco Island and Florianópolis), and the others were recovered on the coast of Rio Grande do Sul (between the Solidão and Verga lighthouses). Two of them were found in the sea: one near Rasa Island and the other one near Alcatrazes Island (Fig. 2). With the exception of these bottles (where the distances covered were approximately 60 nm and 20 nm respectively), the others drifted distances from 240 nm to 680 nm. Only one bottle reached the coast of the studied area (Cabo Frio); all the others took the WSW and SW direction, their velocity ranging between 3 to 6 nautical miles per day (nm/day) (Table 1).

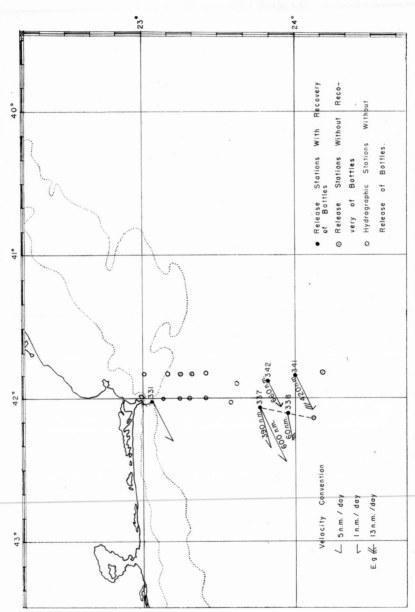

Figure 1 Drift bottle release stations within the study area, showing area distribution of stations from which recoveries were obtained, July 1968

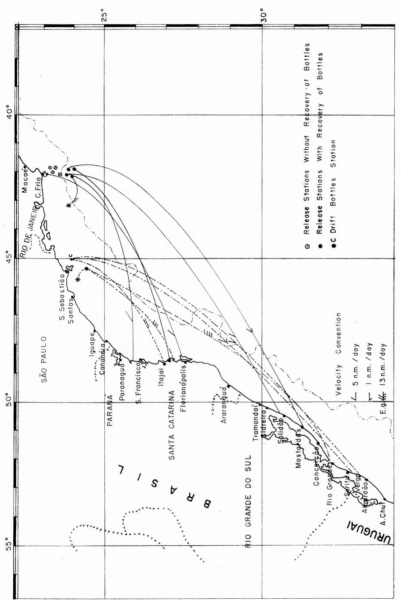

Figure 2 Positions of the drift bottle stations; their areas of recovery, drift paths, and velocities of cruise, July 1968

Table 1 Drift bottles recovered from cruise July 1968

	RELEASE				RECOVERY						
Station Number	Date 1968	Position Latitude ° 'S	Longitude ° 'W		Locality Where Found	Latitude ° 'S	Longitude ° 'W	Date 1968	Probab. days at sea	Distance nm	Calculated Velocity nm/24 h
331	26/7	23 04	042 00		Cabo Frio-RJ	22 51	041 59	20/8	25	13	0.5
					Florianopolis SC	27 25	048 28	29/10	95	428	4.5
337	26/7	23 46	042 04		SE Guaratuba-SC	25 55	048 34	13/10	79	390	4.9
					S Of Cidreira Lgthse-RS	30 30	050 22	16/11	113	600	5.3
338	27/7	23 57	042 00		S Of Rasa Island-RS	23 50	043 09	16/8	20	60	3.0
341	27/7	24 00	041 50		At sea; near-Pôrto Belo-SC	27 06?	048 36?	12/8	16	400?	?*
342	27/7	23 48	041 52		Mostardas Lgthse-RS	31 13	050 49	20/11	116	660	5.7
C+	29/7	23 51	045 02		Near Sarita Lgthse-RS	32 44	052 27	20/11	114	665	5.8
	30/7				Verga Lgthse-RS	32 58	052 34	6/12	129	675	5.2
					Itajaí-SC	26 56	048 38	12/10	74	270	3.6

Table 1 (*cont.*)

Station Number	RELEASE Date 1968	Position Latitude ° ′S	Position Longitude ° ′W	RECOVERY Locality Where Found	Latitude ° ′S	Longitude ° ′W	Date 1968	Probab. days at sea	Distance nm	Calculated Velocity nm/24 h
353	30/7	24 28	045 21	At sea SE Alcatrazes I.-SP	24 10?	045 45?	31/7	1	20?	20?*
				Pôrto Belo-SC	27 06	048 36	13/10	74	240	3.2
				S. José do Norte-RS	32 06	052 02	29/12	151	580	3.8

+C = drift bottles station only
 * = data doubtful

Abbreviations used for Brazilian States:

RJ = Rio de Janeiro
GB = Guanabara
SP = São Paulo
PR = Paraná
SC = Santa Catarina
RS = Rio Grande do Sul

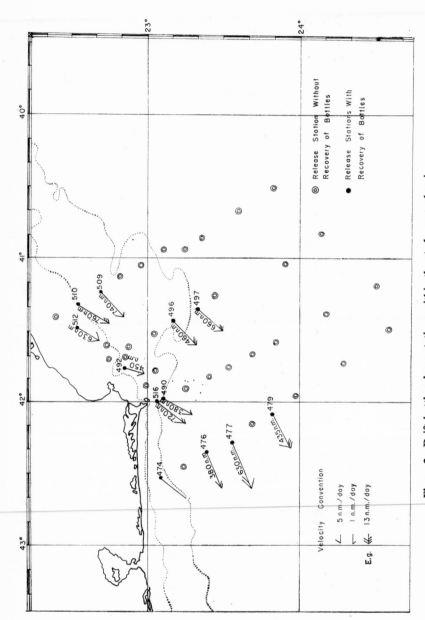

Figure 3 Drift bottle release stations within the study area, showing area distribution of stations from which recoveries were obtained

Table 2   Drift bottles recovered from cruise January 1969

Station Number	RELEASE				RECOVERY						
	Date 1969	Position Latitude ° ' S	Longitude ° ' W		Locality Where Found	Latitude ° ' S	Longitude ° ' W	Date 1969	Probab. days at sea	Distance nm	Calculated Velocity nm/day
474	17/I	23 09	042 32		At Sea-90 nm from I. Grande-RJ	?	?	26/I	9	—	—
476	17/I	23 23	042 21		S. Francisco do Sul-SC	26 14	048 28	13/3	55	380	6.9
477	17/I	23 33	042 17		Solidão Lgthse-RS	30 42	050 29	28/2	42	630	15.0
479	17/I	23 49	042 05		Florianópolis-SC	27 48	048 30	12/3	54	435	8.0
490	19/I	23 06	041 59		Paranaguá-PR	25 34	048 20	29/3	69	380	5.5
492	19/I	22 51	041 46		Florianópolis-SC	27 27	048 22	5/4	76	450	5.9
496	19/I	23 10	041 26		Pôrto Belo-SC	27 10	048 30	17/3	57	461	8.1
497	19/I	23 19	041 21		Solidão Lgthse-RS	30 44	050 30	7/4	78	660	8.5
509	21/I	22 41	041 13		Mostardas Lgthse-RS*	31 15	050 54	3/6	133	740	5.6
510	21/I	22 32	041 19		Capão Comprido-RS	31 25	051 02	22/4	91	760	8.4
512	21/I	22 32	041 29		Capão da Canao-RS	29 46	050 01	30/3	66	630	9.5
516	21/I	23 03	042 00		Camboriú-SC	27 00	048 36	5/3	43	434	10.1
					Bojuru-RS	31 30	051 07	28/3	66	720	10.9

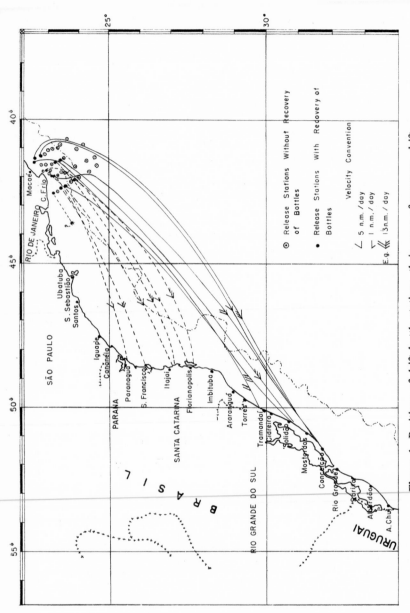

Figure 4 Positions of drift bottle stations, their areas of recovery, drift paths, and velocities of cruise, January 1969

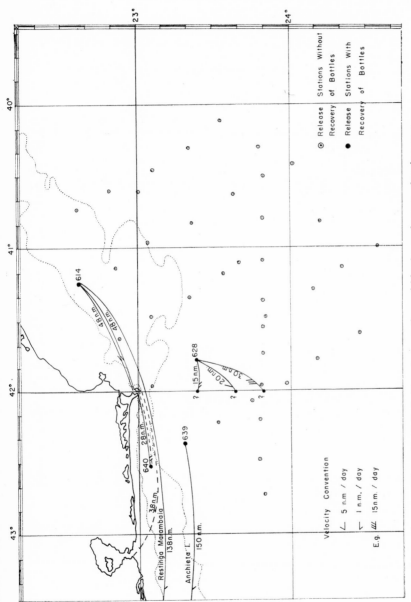

Figure 5 Drift bottle release stations within the study area, showing area distribution of stations from which recoveries were obtained, May 1969

Fertility of the Sea

Table 3 Drift bottles recovered from cruise May 1969

	RELEASE			RECOVERY						
Station Number	Date 1969	Position Latitude ° ′ S	Longitude ° ′ W	Locality Where Found	Position Latitude ° ′ S	Longitude ° ′ W	Date 1969	Probab. days at sea	Distance nm	Calculated Velocity nm/day
614	19/5	22 36	041 15	Cabo Frio-RJ	22 54	042 02	25/6	37	48	1.3
				Cabo Frio-RJ	22 54	042 02	26/6	38	48	1.3
				Rest. Marambaia-RJ	23 92	043 35	28/8	101	138	1.4
628	21/5	23 22	041 47	At Sea Cabo Frio-RJ	23 48?	042 00?	23/5	2	30?	15?
				At Sea Cabo Frio-RJ	23 38?	042 00?	25/5	4	20?	5?
				At Sea Cabo Frio-RJ	23 24?	042 00?	30/5	9	15?	1.7?
639	22/5	23 18	042 22	Anchieta I.-RJ	23 32	045 04	22/9	123	150	1.2
640	22/5	23 04	042 31	Cabo-Frio-RJ	22 58	042 01	7/6	16	30	1.9
				Guanabara-RJ	23 00	043 17	2/11	164	38	0.2?

January 1969

During the oceanographic cruise in January 1969 420 drift bottles were released and 12 (2.8%) were recovered. Half of these bottles were found on the coast of Paraná and Sta. Catarina States (between Paranagúa and Florianópolis); the others drifted further south to the coast of Rio Grande do Sul (between Tramandaí and Conceiçao lighthouses) (Fig. 3). The probable distances covered by the bottles range from 380 nm to 760 nm, the calculated velocities being from 6 to 15 nm/day (Table 2).

None of the recovered bottles reached the coast within the area of survey, as all traveled great distances to the south. The launching points for all the bottles successfully recovered were grouped within the region of the continental shelf. Those drift bottle stations further offshore did not yield any recoveries (Fig. 4).

May 1969

In May 1969, 490 drift bottles were launched, out of which 9 were returned (1.8%). In contrast to the two earlier cruises, some of the bottles were found inside the studied area on the coast and at sea. The exceptions were three bottles which drifted westward, one to Anchieta Island, the other nearshore to Restinga da Marambaia, the third to Guanabara (Fig. 5). The main directions were WSW and SW. Of special interest are 3 bottles launched on the same station which were found at sea. Again no bottles were recovered from stations located outside the shelf region.

All the distances covered were small; only two attained distances of 150 nm and the rest were less than 50 nm. The calculated velocities ranged from 1.3 to 1.9 nm/day (Table 3).

DISCUSSION AND COMPARISON

Two facts are evident from the results: nearly all the bottles ran south-west or west. Those that traveled to the south-west, did so at relatively high speeds (ca. 6 nm/day). The primary mechanism of transport, therefore, must be the Brazil Current.

Yet, bottles actually released in the Brazil Current did not return to the coast, as also observed on numerous other cruises. Furthermore, there is a considerable seasonal variation in the percentage returned: viz. July 1968 (8.7%), January 1969 (2.8%) and May 1969 (1.8%). An explanation for this is that the drift bottles are also subject to the influence of near surface wind-drift currents (Ekman, 1905).

Another interesting fact is that no recoveries occur further south than Rio Grande do Sul, indicating that at this latitude the effect of the Brazil Current is diminished and other currents become more important.

In two cruises (July 1968 and January 1969) great distances were traveled by drift bottles running to the South-West. But the release points for the two cruises were very different. In July 1968 the bottles recovered were released from off-shore regions, that is to say sea-wards of the Brazil Current. In January 1969 the opposite was true and all the recoveries were from the landward side of the Brazil Current. But, in addition the wind-fields were diametrically opposed; the winds blowing from the south and south-west during July 1968 and from the Northeast during January 1969.

To explain these results, a simple "conveyor-belt" mechanism is proposed:

The bottles are carried into the Brazil Current by wind-drift currents; once in the current, they are retained by some dynamic process and are carried rapidly to the south, until conditions favour their expulsion, where upon they are washed up on the coast.

Thus in January 1969 the bottles were driven sea-wards by the NE winds, until they encountered the Brazil Current; and those already outside the inner boundary of the Brazil Current were driven outwards and lost in the centre of the South Atlantic gyre (Sverdrup, 1946).

In July 1968, however, the bottles were driven landwards until they encountered the Brazil Current. The fate of those released on the coastal side of the Brazil Current, is a subject for speculation. One assumes that they entered the coastal current system and were carried eastwards (Mascarenhas et al., 1969). It is possible that this current sporadically penetrates the Brazil Current, consequently also expelling the bottles to the central South Atlantic.

The results of the May 1969 cruise are somewhat anomalous in this simplified scheme. Under the influence of NE winds, there were no recoveries from those bottles launched outside the Brazil Current; nevertheless, in contradistinction the region inside the Brazil Current appeared to indicate a coastal convergence of surface waters which seems to have been responsible for the absence of upwelling during this cruise. (The velocities ranged from 1.3–1.9 nm/day which appears to be typical of wind-drift currents.)

It must be remembered however that bottles may have entered the Brazil Current, but favorable conditions for exit were not encountered in regions further south; or, that bottles exited from the main current in the general area of Santos and then returned, traveling NE in the coastal current system. In particular, such appears to be the case for a bottle that arrived at Rio de Janeiro (only 40 nm distant) after 164 days at sea. A similar situation was also observed in a drift bottle survey of the Rio Grande do Sul region (Luedemann, 1969).

Another question to be asked is the type of mechanism that retains the bottles in the main stream. The possibilities are:

a) Horizontal "line" convergence of surface waters in the main current which could be the effect of wind-produced downwelling at the current boundary.

b) Horizontal shearing effects at the current boundary.

c) Wind-drift currents are not sufficiently strong to take a bottle right across the main current before adverse wind conditions are met.

However, condition "c" is not regarded as probable.

To further support the "conveyor-belt" theory, evidence is drawn from the grouping of the results on the Santa Catarina coast (Lat. 27° S) and the coast of Rio Grande do Sul (Lat. 32° S). Some two-hundred (200) nautical miles of coastline, centered on Cape St. Marta (28° 40′ S) where recovery is minimal, separate these two groups. The group that traveled the greatest distance had the highest apparent drift velocities (Table 4).

We will dry and explain this difference by assuming the times (T in days) for the bottles to enter and leave the main stream are the same for both

Table 4 Group recoveries on the coasts of Santa Catarina and Rio Grande do Sul

Release region	Region Found				Date
	Santa Caterina		Rio Grande do Sul		
	Veloc. nm/day	Distance nm	Veloc. nm/day	Distance nm	
Cabo Frio	7.4	420	9.7	690	Jan. 1969
	4.7	410	5.5	630	July 1968
São Sebastião	3.4	250	4.9	600	July 1968

groups. The difference in apparent velocities is consequently due to the different lengths of time spent in the fast flowing main-stream. Accordingly, we can set up two simultaneous equations for the two groups (Santa Catarina and Rio Grande do Sul) of the following form:

(Total distance run/velocity of the Brazil Current (V))

+ Entry and exit time (T) = Total time spent at sea.

Solving the equations for the two independent variables V and T, we have for January 1969 that the velocity of the Brazil Current $V = 19$ nm/day and the entry-exit time $T = 35$ days. Applying the same technique to the results of July 1968 (winter) gives $V = 8.1$ nm/day and $T = 36$ days, which is a remarkable agreement between the entry-exit times in different seasons of the year.

Further corroboration is obtained from a third set of results, of the July 1968 cruise, from bottles that were launched outside the study area, some 150 miles further south. These give: $V = 7.0$ nm/day and $T = 37$ days.

It must be emphasized that because of the relatively small sizes of the samples the results can only represent an approximation to the real state of affairs. Nevertheless, the results are sufficiently encouraging to merit further investigation.

The width of the continental shelf is approximately 60–80 nm in this region (Santa Catarina and Rio Grande do Sul). Thus the entry-exit time (T) of 35 days would give wind-drift currents of the order of 2 nm/day which is an acceptable result.

One fact deserves a special mention. No bottles have been recovered after any extensive period (>170 days) at sea. Such a fact would appear to indicate that after prolonged exposure to sun and sea-water the sealing-wax weakens and allows marine borers to penetrate the cork. The process is probably hastened, if the sand ballast shifts during rough weather, so that bottle does not float properly. Two improvements are suggested: 1) using a plastic seal over the cork; 2) fixing the sand ballast with gelatine to prevent shifting.

CONCLUSIONS

The influence of the Brazil Current is the primary factor affecting drift bottle returns on the SW coast of Brazil.

Whether a bottle will enter the current or not is determined principally by the direction of the predominant wind-field. In such a manner drift bottles launched at a sufficient number of stations help to delineate the main current boundary. The combined entry and exit times to the main current appear to have an order of magnitude of between 35–40 days. The wind field also appears to have a controlling influence on the exit of drift bottles from the main stream.

There is a great seasonal variation in the Brazil Current. Mean velocity in the winter months between Cabo Frio (Lat. 23° 50′ S) and Rio Grande do Sul (Lat. 30° S) is approximately 1/3 knot; but during the summer months it attains more than double this speed.

Acknowledgements

The authors would like to express their thanks to Mr. L. B. Miranda and Dr. Argeo Magliocca for their helpful suggestions and criticisms.

Thanks are also due to all the personnel of the Instituto Oceanográfico involved in preparing and launching the drift bottles as well as to all the staff that assisted in the preparation of this manuscript.

A special thank you is offered to Mr. W. de C. Dotti, who very kindly helped with the preparation of the drawings.

References

Brasil. Directoria de Hidrografia e Navegação—(1954) *Lista de Faróis*. (1968) Roteiro

EKMAN, V. W. (1905). On the influence of the earth's rotation on ocean currents. *Ark. f. Mat. Astron. Och Fysik*, **2**, 11.

LUEDEMANN, E. F. (1967). Preliminary results of drift bottle releases and recoveries in the Western Tropical Atlantic. *Bolm. Inst. Oceanogr. S. Paulo*. **16**, 1, 13–22.

LUEDEMANN, E. F. (1969). Relatório sóbre resultados obtidos com lamçamentos de garrafas de deriva realizados durante o "Programa Rio Grande do Sul"—Presented to the group GEDIPE. *Sáo Paulo, Inst. Oceanogr. de USP*. vol. **2**, 22 p.

MAGLIOCCA, A., and LUEDEMANN, E. F. (1969). List of recovered drift bodies (1955 to 1963). *Contrçoes Inst. Oceanogr. USP-Sér. Ocean Fís*. no **12** (in press).

MASCARENHAS, A. S., MIRANDA, L. B., and ROCK, N. J. (1969). A study of the oceanographic conditions in the region of Cabo Frio. Paper presented at the *International Symposium of "Fertility of the Sea"*. S. Paulo, 1–6 Dec. 1969.

NEUMANN, H. (1968). Die Trift von Verschmutzungen an der Oberfläche der Nordsee. (The drift of pollutants at the surface of the Nord Sea). *Helgolaender wiss. Meeresunters*. **17**, 81–93.

SVERDRUP, H. U., JOHNSON, N. W., and FLEMING, R. H. (1963). *The Oceans*. New Jersey, Prentice Hall, Inc. 1078 p.

A study of the oceanographic conditions in the region of Cabo Frio

A. S. MASCARENHAS, JR., L. B. MIRANDA
and
N. J. ROCK

Instituto Oceanográfico da Universidade de Sao Paulo, Brazil

Resumo

A análise dos resultados obtidos durante os cruzeiros oceanográficos realizados no período de janeiro de 1968 a julho de 1969, mostram a ocorrência de ressurgência de águas frias e pouco salinas na região de Cabo Frio, a distância da costa inferior a 10 milhas náuticas. Verificouse que o fenômeno ocureu com maior intensidade em janeiro de 1968 a julho de 1969 na região situada á oeste de Cabo Frio.

Observações meteorológicas efetuadas à bordo, evidenciaram que ventos variando de NNE a ENE foram os mais efetivos em produzir correntes superficiais divérgentes da costa, as quais, durante os meses de janeiro e maio de 1969, não foram compensados por advecção vertical (ressurgência).

Um modêlo hipotético da provável estrutura oceanográfica, induzida por ventos de NE é sugerida, a fim de, explicar a grande variedade de condições de ressurgência encontradas na região em estudo.

Physical oceanographic research in Brazil dates from 1956, when the first attempts to describe the regional oceanography were carried out by the Instituto Oceanográfico (University of Sao Paulo) and the Diretoria de Hidrografia e Navegaçao (Brazilian Navy).

Emilsson (1956) reported upwelling for the first time to the south of Cabo Frio, evidenced by a strong horizontal temperature gradient at the sea surface. Later the Brazilian Navy, as part of the IGY 1957 program, reported upwelling south of Cabo Frio and also on the coast of Espírito Santo (north of Cabo Frio) (D.H.N. Cruise reports 1957).

Subsequently, Emilsson (1961) wrote the most complete treatise of Brazilian coastal waters known to date, based mainly on oceanographic observations of 1956 in the region between the Abrolhos Banks (lat. 18° S) and the River Plate estuary (lat. 35° S). In this work a qualitative theory was proposed, suggesting that upwelling is maintained by a thermohaline process.

A more recent publication by Moreira da Silva *et al.* (1966), which was based on a ten-day time-series of hydrographic observations from Nansen bottle casts, also refers to the process of upwelling. The time series was obtained at an anchor station by the N/Oc. *Almirante Saldanha*, on the edge of the continental shelf slightly to the east of Cabo Frio (22° 55′ S and 40° 52′ W). The above authors found that the changes of the water masses on the shelf seemed to be casually related to changes in the wind field. The changes were ascribed to the process of upwelling. On this basis, and remembering that the conclusions are drawn from water masses at the *edge* of the continental shelf, one must deduce that upwelling occuring at the coast influences the coastal water masses over the whole continental shelf.

On the other hand, the work of Emilsson (1961) emphasizes that upwelling was restricted to an area at the edge of the continental shelf and suggests that the dominant influence cannot be wind stress, but rather a thermohaline mechanism.

Five oceanographic cruises were made by the Oceanographic Institute aboard the N/Oc. *Prof. W. Besnard* in January, and July of 1968, and in January, May and July of 1969 in the Cabo Frio region (Fig. 1).

On each of the cruises standard hydrographic casts were made with Nansen reversing water bottles. Temperature was measured by means of reversing thermometers, and salinity was calculated from conductivity measurements on water samples with an inductive salinometer bridge.

Meteorological data was kindly provided by the chemical firm Companhia Nacional de alcalis for the January 1968 and January 1969 cruises from the company's meteorological station situated at the cape. This was supplemented with meteorological data taken on board ship and marine weather reports from other shipping and weather stations (Villela, 1969).

Current measurements in the area were attempted with parachute drogues and Ekman current meters. Neither gave satisfactory results due to the lack of precise navigational aids in the area. Drift bottles were also released and, although the number of returns was small, significantly aided the interpretation of oceanographic surface conditions.

The analysis of the situation was conducted principally by graphical means, and included:

a) Vertical sections of salinity, temperature and sigma-t.

b) Horizontal distributions of temperature and salinity at the surface (10 m), on surfaces of equal sigma-t (an approximation to isentropic surfaces), and also over the sea-floor of the continental shelf (from such distributions, patterns of circulation and the mixing of water masses over these surfaces were inferred).

c) The dynamic topography off the continental shelf plotted with reference to the 500 and 800 db surfaces.

d) Scatter diagrams of T-S characteristics over selected surfaces.

Cabo Frio is region of transition, both of continental shelf features and of water masses. To the north (the Brazilian east coast region), the continental shelf is narrow and Tropical oceanic water masses predominate, whereas to the south (the south-western region), the continental shelf is broad and water masses on the shelf show a complex interaction of many factors. This transition is clearly expressed within the actual area studied. The continental shelf to the north of Cabo Frio is approximately 55 miles wide, and for the major part is shallower than 100 m depth. Tropical water masses predominate on the shelf, and the continental slope, running in a line N.E., falls away steeply to the abyssal plain (Fig. 2).

In contrast, the shelf region to the south and west of the cape is comprised of a broad plateau (some 40 miles wide), delimited on the outer edge by the 200 m isobath and generally deeper than 100 m. The edge of the shelf (marked by the 200 m isobath) runs E.-W., and the area where the shelf changes direction is marked by a shallow submarine "salient" (sea-rise) from the shelf edge that rises to within 70 m of the sea-surface.

THE 1968 CRUISES

In January and July 1968, four hydrographic sections were made in the region south of Cabo Frio. In Figure 3a the vertical distribution of temperature and salinity in the section completed on January 24 shows evidence of a strong upwelling of cold and less saline waters, occuring very close to the Cape. The surface temperature gradient in this region was very steep, being almost 0.5°C/nm. The subsurface salinity maximum, occuring at depths less than 50 meters, seems to indicate a surface outflow of less saline surface waters in the offshore direction. At this time strong winds were blowing from NNE and ENE, with speeds up to 16 m/sec. during at least 24 hours. Thus it appears that this out flow was wind driven according to Ekman's (1905) theory. A comparison between the vertical TS diagram for the outer-most station (St. 248) and the horizontal TS scatter diagram along the section at different depths shows that the surface coastal waters are lifted from a depth of about 150 meters. Below 50 meters depth, the water masses along the section have the same characteristics as the water masses in the vertical (St. 248) showing that the influence of vertical turbulent mixing is weak during the upwelling process.

Accordingly, we may assume that the main influence changing temperature of the surface water mass is the daily heat input at the surface. Using an average value of 300 ly/day for the daily heat budget, one concludes that the surface water sampled on Station 24 had upwelled at least four days previously.

For the two remaining sections made between 26 and 28 of July there is no evidence of upwelling, although a core of cold, less saline water (temperature of 13°C and salinity 35.10°/$_{00}$) was occupying the bottom layers on the continental shelf close the coast, as indicated by vertical sections (Fig. 3 b).

Classification of bottom water masses

Bottom water masses on the shelf are characterized by two principal groups (Fig. 4), which we will call:

1-Sub-Tropical Shelf Water (STSW) (12°C–15°C; 35.1–35.5°/$_{00}$), and con-stitutes the principal water mass of upwelling. This water-mass has its origins in a layer situated between 200 m–300 m depth off the continental shelf.

2-Tropical Shelf Water (TSW) (17°C–19°C; 35.90–36.30°/$_{00}$) which appears to be formed as an intermediate water-mass between the Sub-Tropical Shelf Water and the waters of the Brazil Current. The latter water mass may be generally described as Tropical Oceanic Surface Water (TOSW) with different characteristics (22°–25°C; 36.60–37.00°/$_{00}$).

THE JANUARY 1969 CRUISE

In January 1969 (the southern summer), the Brazil current was found as a fast narrow current flowing parallel to the eastern edge of the continental shelf (Fig. 5a). It is apparent that the current is sensitive to the 100 m isobath, as on reaching the southern limit of this contour, a branch of the main surface current swings westwards onto the shelf and continues west until the vicinity of Cabo Frio. In the south western part of the shelf, another branch of the Brazil current is found flowing *eastwards* paralleling the 100 m isobath (Fig. 3c). These two branches join about 30 miles S.E. of Cabo Frio, and the main flow turns south and leaves the shelf slightly to the west of the sea-mount described (Fig. 2).

Part of the flow forms a large anti-cyclonic eddy around the sea-mount, and another part contributes to a cyclonic eddy situated on the shelf approximately 50 miles south of Cabo Frio.

Drift bottle returns appear to delineate the inner edge of the Brazil current (Fig. 6a) and thus the outline of the anti-cyclonic eddy. Otherwise, the circulation on the shelf is inferred from horizontal distributions of temperature and salinity (Fig. 5a).

The meteorological situation prevailing for January was typical to provoke upwelling; the mean wind speed being 3.7 m/sec from the N.E. with a relative frequency of 55% and calm 18% of the time. Such a situation provoked a transport of lighter surface water away from the eastern coast, establishing a longshore density or "slope" current running S.W.

Upwelling occurs on this coast but only in the extreme north of the region. This occurs because the S.W. water masses meet eastward flowing coastal currents in the southern part of the region and form a region of confluence and strong mixing which inhibits upwelling.

On the southern coast the windstress has a diminished effect due to the sheltering effect of the coast, and also because the component of windstress parallel to the coast is small. Accordingly, it is deduced from the mixing of surface waters that there exists an eastward flow of coastal waters near to the coast which is due to the semipermanent, horizontal density gradient created by the run-off and coastal outflow of the region.

The interaction of this density-produced flow with the eastward flowing branch of the Brazil current creates a counter-current flowing westward between the two streams which has its origins on the shelf east of Cabo Frio.

In the eastern region of the shelf a new water-mass ($T = 19°C$; $S = 36.2°/_{00}$) appears to form as a result of the mixing of different water masses and flows off the shelf to the east, forming a counter current to the Brazil current at a depth of 100–200 m (Fig. 2).

Quasi-isentropic analysis show that, in order to upwell in the extreme north of the region, Sub-Tropical Shelf Water crosses the 200 m isobath in the eastern part of the continental shelf and flows westward as a broad stream across the plateau of the continental shelf (Fig. 7a). Just south of Cabo Frio, it turns north and, crossing the 100 isobath, flows as a narrow bottom current parallel to the coast. Consequently, the upwelling water in the north is only weakly marked, as after its long course across the shelf it contains more than 65% of Tropical Shelf Water.

THE MAY 1969 CRUISE

In this cruise, five sections at right angles to the coast and one, running along the parallel of 23°50'S, were made. At 67% of the hydrographic stations water depths exceeded 500 meters, and for these stations geostrophic currents computations have been carried out.

According to Mr. Frank R. Shaffer, Geologist of the Oceanographic Institute (personal communication), submarine canyons or valleys are present in the region south of Cabo Frio. To confirm these bottom topographic features, and to sample their water characteristics, the latitudinal section was carried out with closely spaced hydrographic stations and continuous echo-soundings.

Figure 8 shows the detailed vertical distributions of temperature and salinity along the section, as well as the surface variation of these properties. Interesting to note are the topographic features between stations 593 and 598 which indicate a sub-marine valley and peak. Associated with this peak is a surface temperature anomaly of $-2°C$ and surface salinity anomaly of $-0.5°/_{00}$ which are probably due to a warm cyclonic vortex with a faster moving upper layer, situated around the submarine peak. The bottom water found in the valley contains slightly more than 70% of Antarctic Intermediate Water, and some 30% of TOSW.

The results of geostrophic currents are indicated in Figure 6b. The dynamic isobaths are drawn with 0.02 dynamic meter intervals and the inset diagram shows the theoretical relationship between the geostrophic velocity and the distance between contours. It can be seen that the main axis of the Brazil Current is located close to the edge of the continental shelf flowing in the southwest direction, until the mean position of 23°50'S and 41°20'W is reached. Afterwards, it turns abruptly to the south-southeast direction and, crossing the 24° parallel, it turns west. The meandering of the main flow of the Brazil Current appears to be the result of bottom topographic effects, as around st. 600, the bottom rises abruptly in the western direction (Fig. 8).

Evidence of a small anticyclonic vortex is found in the northeast of the studied region and a larger one is found in the south. Vortices in precisely this region have already been reported in the D.H.N. Cruise report of June 1960.

The coastal currents, inferred on the basis of the horizontal temperature and salinity distribution at 10 meters depth, (Fig. 5b) show the following important features:

1.   An eastward current of cold and less saline coastal waters (probably a compensation flow) that converges south of Cabo Frio with a warmer and more saline coastal current running southwards, no doubt a branch of the Brazil Current.

2.   The Brazil Current, which is clearly marked as a ribbon of warm, saline water with temperatures and salinities higher than 25°C and 36.90$^o/_{oo}$ respectively.

The wind system prevailing during this cruise was predominantly from N and ENE, with a relative frequency of about 60%, and speeds ranging from 3.0 to 12.5 m/s. The southward flowing coastal current, as inferred by the horizontal distribution of isolines, deflects to the left when it crosses the southern limit of the 50 m isobath, supporting the supposition that this coastal current is wind driven. It appears that this offshore water transport is compensated by an eastward flowing coastal current (see above) which inhibits the upwelling process.

Salinity and temperature distribution on the 26.0 Sigma-t surface (Fig. 7b), located inside the thermocline at depths on the continental shelf between 40 to 90 metres, ranges from 17.90°C to 19.25°C and 35.90$^o/_{oo}$ to 36.30$^o/_{oo}$ respectively. The tongue-like distribution of both temperature and salinity southeast of Cabo Frio indicates that the eastward flowing coastal current spreads its characteristics in the southeasterly direction by turbulent mixing.

THE JULY 1969 CRUISE

In this cruise, a hydrographic survey of a small sea area S.E. of Cabo Frio Island was made to obtain oceanographic data in support of a flight by a NASA aircraft equipped with remote sensors. The maximum sampling depth seldom exceeded 200 meters in order to assure a rapid, quasi-synoptic coverage of the area.

The data were analysed following the same steps as the preceding cruises, with the exclusion of geostrophic analysis. At the beginning of the cruise, the meteorological conditions were stable with the wind from the N.E. occasionally reaching wind-speeds of 11 m/sec. During the cruise the wind changed direction and commenced to blow from the W. and finally settled, blowing steadily from the S.W.

Analysis of the temperature and salinity distribution at 10 meters depth (Fig. 9a) shows a surface anomaly of cold and less saline waters, situated close to the coast just west of Cabo Frio and clearly indicating upwelling.

The path of the upwelling water ($T = 14.0°C$ and $S = 35.40°/_{00}$) can be clearly traced across the floor of the submarine plateau previously described, and indicates that the upwelling water mass was also present on the eastern shelf, although only weakly developed upwelling existed (Fig. 9b). Such an interpretation is supported by vertical sections made parallel to the coast.

On the outer region of the eastern shelf there exists a broad, south-westerly flow of the Brazil current that flows out over the colder denser STSW in the plateau region, producing a distinctive three-layer structure with the lighter coastal water forming the upper-layer.

The salinity distribution over the 26.0 Sigma-t surface is shown in Figure 9c. In comparison with the previous 1969 cruises, this figure indicates considerable changes in the spreading of the water masses, and a well developed divergence of upwelling water is observed.

The interpretation placed upon the results is that the wind from N.E. creates a classical situation for upwelling in the area west of the cape and also in the northern part of the eastern shelf. The sudden reversal of the wind to a S.W. direction brings about the counter-acting process of downwelling. On the eastern shelf therefore, where the wind direction parallels the coast, upwelling is rapidly suppressed and leaves the upwelling water-mass isolated on the shelf. However, in the Cabo Frio region the upwelling process meets little resistance and, after a period of about 36 hours subsequent to the cessation of N. E. winds, was still strongly in evidence at a distance of 4 nm from the coast. It is believed that this situation was favoured by changes in the coastal circulation due to seasonal differences in the thermal radiation and precipitation-evaporation balance and also the weakening of Brazil Current in winter. Accordingly, favourable wind conditions during a sufficient time brought about upwelling.

CONL USIONS

In the January 1969 cruise there were conditions conducive to upwelling; but, due to the dynamic barrier formed by a branch of the Brazil current flowing close to the coast, it was effectively inhibited, and upwelling occurred only on the eastern coast. Upwelling was present, however, in the center of an warm anticyclonic eddy, shown in Figure 3c, a situation which the present authors believe to be analogous to those discovered by both Emilsson (1961) and Moreira da Silva *et al.* (1966).

During May 1969, upwelling was not observed. A distinct absence of coastal water was noted, and Tropical Water masses had invaded the shelf regions in both the East and the South. Although prevailing wind conditions from the N.E. were sufficient to provoke upwelling on the east coast, no marked area of upwelling existed due to the intrusions of Brazil current waters on the shelf and because the bottom waters of this shelf region were wholly Tropical. In the southern area of the east coast, divergence of surface waters was horizontally compensated by a eastward coastal current which is apparently confirmed by drift bottles recoveries. It is possible that upwelling existed on the south coast at distances from the coast less than 8 nm but was not taken in by the station network.

Concluding the presentation, a tentative model for the circulation of the Cabo Frio region may be outlined from the data. Winds from the N.E. appear to provoke the passage of upwelling water across the eastern part of the plateau region. In the region of Cabo Frio a portion upwells, but another portion turns northwards to upwell on the northern sector of the coast. A combination of different factors (e.g. the changes in the wind system, oscillations of the main axis, and strength of the Brazil Current, or compensation by horizontal coastal currents) alters the situation to produce a variety of different results.

19a*

A semi-permanent eastward flowing coastal current is generally present on the south coast, and coastal water mixes with the southward flowing, wind-produced current on the eastern shelf to form a new water mass which spreads out in a southeasterly direction. There seems to be enough evidence that at deeper levels this water mass is the source of a counter-current to the Brazil Current.

The topographic features of the sea-floor in the Cabo Frio region act greatly to influence the circulation pattern and appear to be a direct factor in inducing eddies.

Certain facts remain unexplained: a cold core of Sub-Tropical water always persists on the shelf in all seasons of the year, even though upwelling is not present. Is this due to the thermohaline mechanism proposed by Emilsson or perhaps a "pumping" mechanism produced by dislocations of the Brazil Current? To what extent do the submarine valleys and canyons facilitate the upwelling process, perhaps by channeling the upwelling waters?

Finally, what are the time scales of the upwelling, the volume transport, and the magnitude of the dynamic factors involved? All these questions must await modern equipment before they can be properly answered.

Acknowledgements

The authors wish to express their thanks to all members of the Instituto Oceanográfico who made this publication possible. In particular special thanks are due to Mrs. Ellen Fortlage Luedemann for her work in drift bottles, Mr. Rubens Junqueira Villela for the preparation of Meteorological reports and Dr. Argeo Magliocca for his discussion and criticism.

The latter author wishes to acknowledge that the work was carried out while he was contracted as a Visiting Professor by the Ford Foundation.

References

EKMAN, V. W. (1905). On the influence of the earth's rotation on ocean currents. *Ark. f. Mat. Astron. Och Fysik*, **2**, No. 11.

EMILSSON, I. (1956). Relatório e resultados físico-químicos de três cruzeiros oceanográficos em 1956. *Contr. avul. do Inst. Oceanogr. da U.S.P.* No **1**.

EMILSSON, I. (1961). The shelf and coastal waters off southern Brazil. *Bol. do Inst. Oceanogr. da U.S.P.* Tomo XI-fasc. **2**.

Brasil. Miniéstrio da Marinha (1957). Relatório dos Cruzeiros Oceanográficos do NE Almirante Saldanha. Publicações DG-06-II, e DE-06-III, D.H.N.

Brasil. Ministério da Marinha (1960). Estudo das condições Oceanográficas entre Cabo Frio e Vitória, durante o outono (Abril–Maio). Publicação DG-06-X, D.H.N.

MOREIRA DA SILVA, P. C., and RODRIGUES, R. F. (1966). Modificações na estrutura vertical das águas sôbre a borda da plataforma continental por influencia do vento. *I. Pq.M. Nota Técnica* no 35/1966.

VILLELA, R. J. (1969). Resultados Meteorológicos dos Cruzeiros 1/69, 4/69 e 5/69 do N/Oc. *Prof. W. Besnard* a Cabo Frio (unpublished manuscripts).

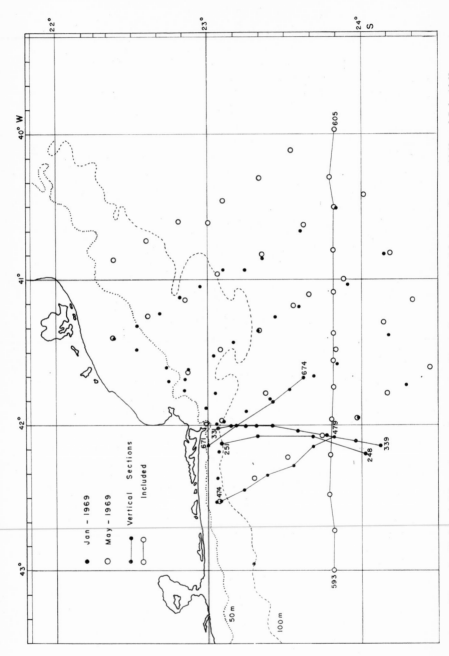

Figure 1 Hydrographic stations and oceanographic sections completed between January 1968 and July 1969

Figure 2 Schematic circulation off Cabo Frio in January 1969. Full lines—surface currents; dashed lines—bottom currents

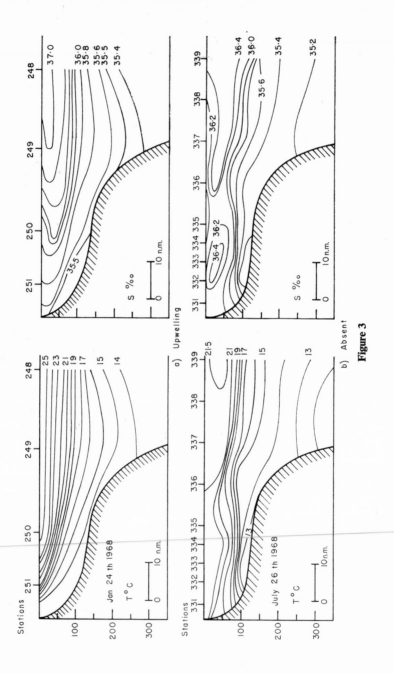

Figure 3

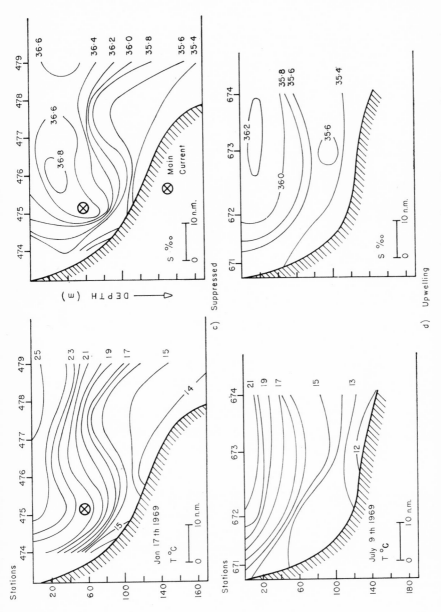

Figure 3 Vertical sections west of Cabo Frio showing conditions of upwelling present and absent

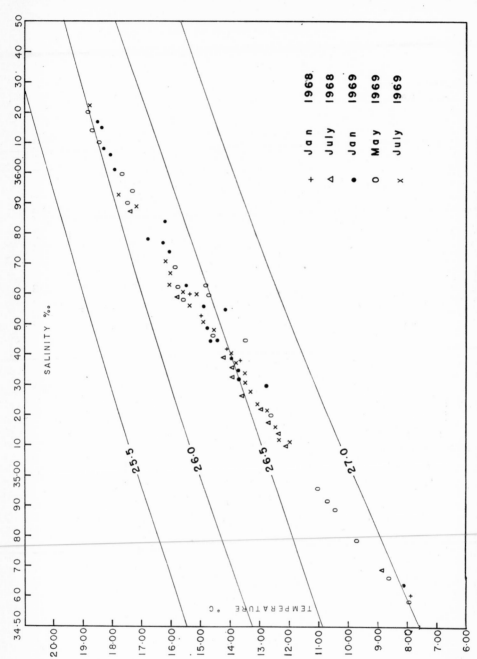

Figure 4   TS characteristics of bottom waters over the continental shelf between January 1968 and July 1969

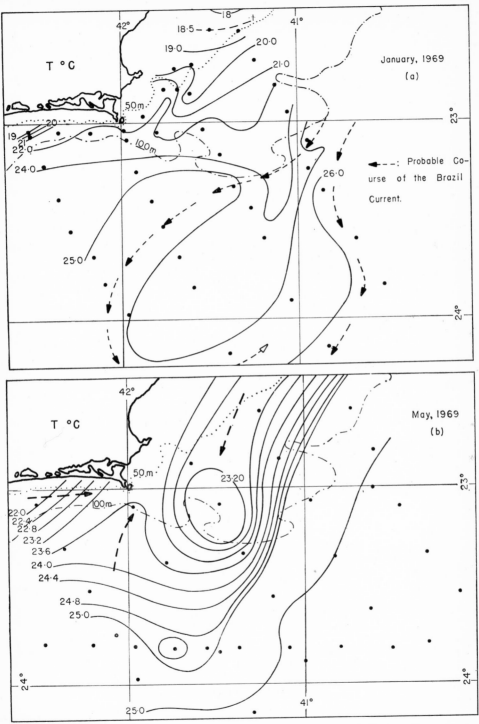

Figure 5 Horizontal distribution of temperature at 10 m depth. Arrows indicate surface currents

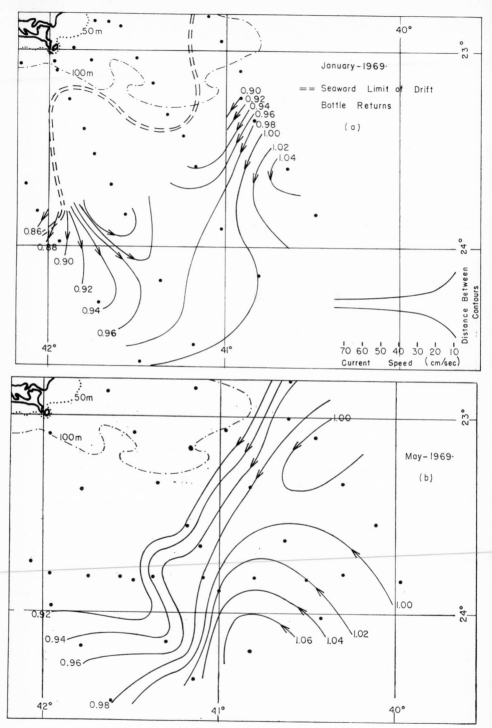

Figure 6 Dynamic topographies of the sea surface relative to the 500 decibar surface

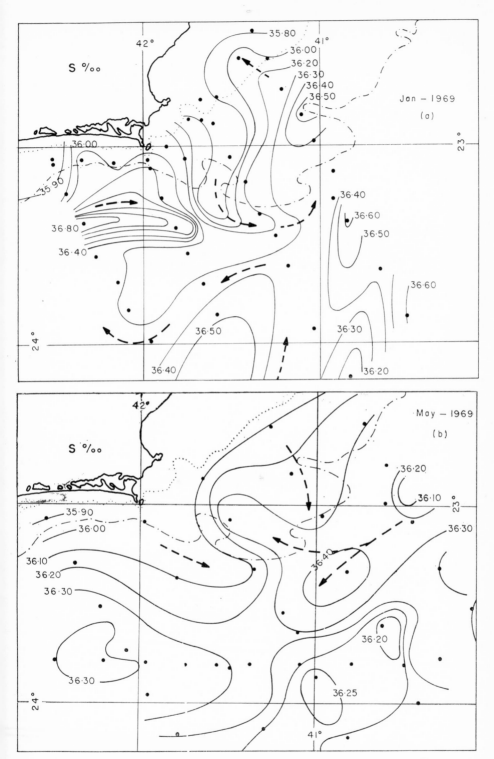

Figure 7 Salinity distribution over the 26.0 sigma-T surface. Arrows indicate spreading patterns

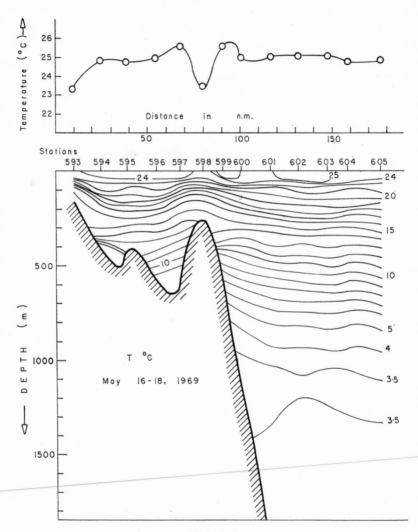

Figure 8-1

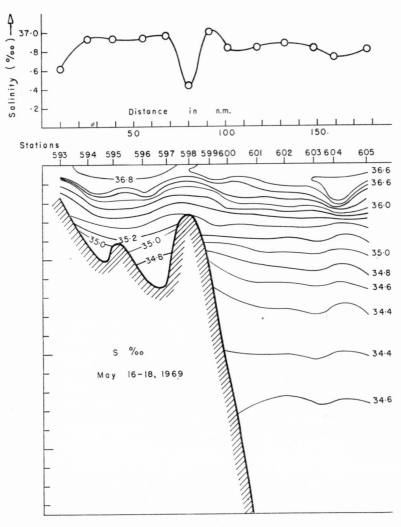

Figure 8-2

Figure 8 Vertical sections of temperature and salinity along the 23° 50′ S parallel. Upper graph shows the surface variation of temperature and salinity along the section

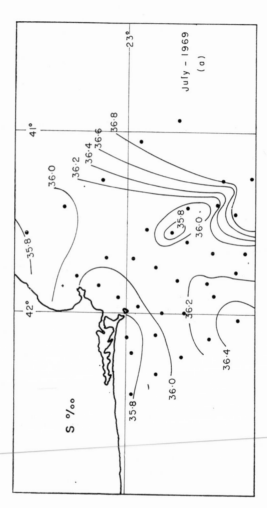

Figure 9a

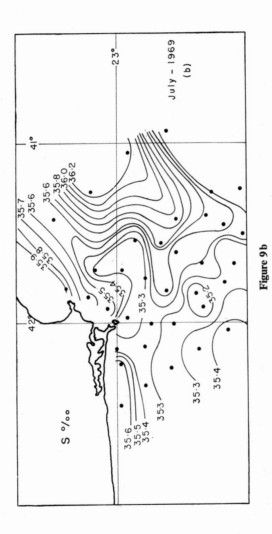

Figure 9b

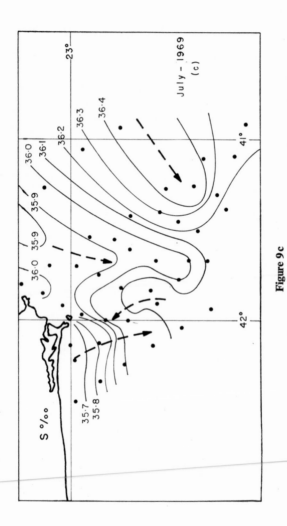

Figure 9c

Figure 9 (a) Horizontal salinity distribution at 10 m depth, (b) horizontal distribution of bottom salinities over the continental shelf, and (c) salinity distribution over the 26.0 sigma-T surface